T0298676

pH-RESPONSIVE MEMBRANES

pH-RESPONSIVE MEMBRANES

Biomedical Applications

Randeep Singh, Piyal Mondal, and
Mihir Kumar Purkait

CRC Press
Taylor & Francis Group
Boca Raton London New York

CRC Press is an imprint of the
Taylor & Francis Group, an **informa** business

First edition published 2022
by CRC Press
6000 Broken Sound Parkway NW, Suite 300, Boca Raton, FL 33487-2742

and by CRC Press
2 Park Square, Milton Park, Abingdon, Oxon, OX14 4RN

© 2022 Randeep Singh, Piyal Mondal and Mihir Kumar Purkait

CRC Press is an imprint of Taylor & Francis Group, LLC

ISBN: 978-1-032-06167-2 (hbk)
ISBN: 978-1-032-06168-9 (pbk)
ISBN: 978-1-003-20101-4 (ebk)

DOI: 10.1201/9781003201014

Typeset in Times
by MPS Limited, Dehradun

Contents

Preface

The synthesis of pH-responsive membranes and their applications in the biomedical field is a new area of study focused on by researchers. The stimuli-responsive membranes represent a new branch of membrane science. Recently, there is a lot of interest and enthusiasm seen in this field. Membrane scientists across the world are involved in the development of different stimuli-responsive membranes for diverse applications in various fields. The most important is the use of stimuli-responsive membranes, especially pH-responsive membranes for biomedical applications, such as drug delivery, hemodialysis, chemical sensing, etc. Therefore, this book gives a short overview of the pH-responsive membranes, by focusing on their biomedical applications. Further, this book is suitable to address the queries of undergraduate, post-graduate, and graduate-level students.

The motivation to write this book is to provide a single, focused, and dedicated source of biomedical applications of pH-responsive membranes. The potential research carried out in the area, as well as the future scope, is discussed, which will provide better insights to the readers about the subject ranging from the basics to the advanced level of the topic. This book covers the initial development and recent use of pH-responsive membranes for various biomedical applications, all in very focused form. It can be used as a stepping-stone for readers going into the field of stimuli-responsive membrane science.

This book is summarized with the following contents:

- Covers a very new topic in the field of membrane science
- Delivers focused knowledge on the topic
- Addresses current world biomedical problems
- Provides insights for the development of new materials and membranes to address important biomedical issues
- Includes all of the major biomedical applications of pH-responsive membranes
- Discusses future insights about applications of pH-responsive membranes

Authors

Dr. Randeep Singh is presently a postdoctoral researcher in Hanyang University (ERICA Campus), Ansan, Republic of Korea working in the field of semiconductor processing. Before joining Hanyang University, he earned his BTech degree in biotechnology from Kurukshetra University, Kurukshetra, Haryana, India (2011); MTech degree in chemical engineering (2013) from the National Institute of Technology Trichy, Tiruchirappalli, Tamil Nadu, India; and PhD in chemical engineering (2019) from the Indian Institute of Technology Guwahati, Guwahati, Assam, India. He was a visiting research fellow at National Taipei University of Technology, Taipei, Taiwan (2018-2019) under the TEEP (Taiwan Experience Education Program) fellowship for his postdoctoral research. His areas of research are in the fields of membrane science and technology, smart materials, nanocomposites, adsorption, biofuels, sonochemistry, desallination, wastewater treatment, and semiconductor processing. His research work is dedicated to the synthesis of various novel materials and membranes with different biological and environmental applications. Presently, he has published 10 peer-reviewed journal articles, 3 books, and 5 book chapters in the fields of materials and membrane science with more than 300 Google Scholar Citations. Further, he presented his research at various national and international conferences. He has received several awards in his field including the Young Scientist Award at the International Science Congress (ISC-2015) held in Nepal. He is an associate member of the Indian Institute of Chemical Engineers and a frequent reviewer for numerous journals related to his research fields.

Dr. Piyal Mondal earned his BTech in chemical engineering from the National Institute of Technology Durgapur, West Bengal (India) in 2012. He completed his Master's degree (2015) and PhD (2021) in chemical engineering from the Indian Institute of Technology Guwahati (India). His research work is dedicated to preparing various surface-engineered polymers for specific environmental applications. Synthesis of polymeric membranes, green synthesized nano-materials, and hybrid techniques to advance wastewater treatment are also his research foci. He has fabricated different prototypes for environmental separation applications. Currently, he has co-authored four reference books: *Stimuli-responsive Polymeric Membranes* (Elsevier, ISBN: 9780128139615), *Treatment of Industrial Effluents* (CRC Press, ISBN: 9780429401763), *Thermal Induced Membrane Separation Processes* (Elsevier, ISBN: 9780128188019), *Hazards and Safety in Process Industries: Case Studies* (CRC Press, ISBN: 9780367516512), with a few more in progress. Moreover, his publications consist of 16 peer-reviewed articles in reputed international journals, with several more under review. He has presented more than 15 papers and received several awards in poster and paper presentations in his field at international and national conferences.

Dr. Mihir Kumar Purkait is a professor in the Department of Chemical Engineering and dean of the Alumni and External Relation office at the Indian Institute of

Technology Guwahati (IITG). Prior to joining the faculty in IITG (2004), he earned his PhD and M Tech in chemical engineering from the Indian Institute of Technology, Kharagpur (IITKGP) after completing his B Tech and B Sc (Hons) in chemistry from the University of Calcutta. He has received several awards, such as Dr. A.V. Rama Rao Foundation's Best PhD Thesis and Research Award in Chemical Engineering from IIChE (2007), BOYSCAST Fellow Award (2009–10) from the DST, Young Engineers Award in the field of Chemical Engineering from the Institute of Engineers (India, 2009), and Young Scientist Medal from the Indian National Science Academy (INSA, 2009). Prof. Purkait is a fellow of the Royal Society of Chemistry (FRSC) UK, and a fellow of the Institute of Engineers (FIE) India. He is also a technical advisor of Gammon India Ltd and Indian Oil Corporation, Bethkuchi for their treatment plant. His current research activities are focused in four distinct areas: (i) advanced separation technologies, (ii) waste to energy, (iii) smart materials for various applications, and (iv) process intensification. In each of the areas, his goal is to synthesize stimuli-responsive materials and to develop a more fundamental understanding of the factors governing the performance of the chemical and biochemical processes. He has more than 20 years of experience in academics and research and published more than 200 papers in different reputed journals and has authored 6 books (Citation: >11700, h-index = 61, 10 index = 142). He has 8 patents and has completed 24 sponsored and consultancy projects from various funding agencies. Prof. Purkait has guided 18 PhD students.

1 Introduction to Responsive Membranes

1.1 INTRODUCTION

Responsive membranes respond to an external stimuli by inducing change in their structural and functional properties [1–3]. This responsive behavior adds to their separation and permeation properties. In general, pH, temperature, light, electric, magnetic, biological, or chemical are some of the most common examples of an external stimuli. Responsive membranes add a new dimension to the field of membrane science. This development increases the potential of membrane science in various fields, such as healthcare, biotechnology, environment, and energy. The responsive membranes, due to their properties of change in structure and function, are developed and explored positively for controlled drug delivery, gas and water treatment, sensing, and antifouling applications [2].

Membranes are made responsive to an external stimulus by using various stimuli-responsive groups. The stimuli-responsive group can either be blended or grafted/coated over the membrane surface so as to make the membrane a responsive membrane. However, fundamental concept mechanisms need to be understood in depth for better utilization of stimuli-responsiveness in membranes. The responsive membranes show a functional and/or structural response, or both, like ionic polymers, shape memory polymers, or shrinkable polymer brushes, respectively. Further, the external stimuli, such as pH, temperature, biological, light, etc., are bulk stimuli. The bulk stimuli is inefficient in the way that it is present in the system even when it is not required; this results in the waste of energy as well as resources. For example, in the case of controlled drug delivery on arrival of a particular foreign body, the drug should be released. Conversely, due to the bulk stimulus, the drug is released in more than the required amount as well as where it is not required and this is how its efficiency and effectiveness both become affected. The local stimulus is far more efficient and effective compared to a bulk stimulus. Hence, a local stimulus is required to be developed for better results.

The external stimuli are categorized into three categories, as shown in Figure 1.1, namely direct, indirect, and field, for their better understanding [4]. In case of direct stimulation, the stimuli groups are in direct contact with the membranes. On the other hand, the responsive sites present on the membrane surface respond to the external stimuli. This is the case where the stimulus is a thermodynamic environment in itself; for example, temperature or pressure. Lastly, in the case of the field stimulation, the membranes respond to an external field, such as electric, magnetic, or electromagnetic.

DOI: 10.1201/9781003201014-1

1

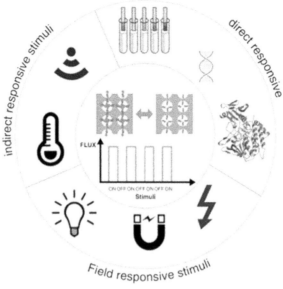

FLUX

ON OFF ON OFF ON OFF ON
Stimuli

indirect responsive stimuli

direct responsive

Field responsive stimuli

Current Opinion in Chemical Engineering

FIGURE 1.1 The external stimuli that modulate membrane performance divided into three groups (reproduced with permission from Darvishmanesh et al. [4] © Elsevier).

1.2 TYPES OF RESPONSIVE MEMBRANES

The recent development and advancement in membranes diversify the field of membrane science. Nowadays, membranes are used on a large scale for various applications and this is due to the wide applicability of the membranes. Membranes are used in wastewater treatment plants, desalination plants, osmotic power plants, sensors, and different types of filters [3]. The inherit membrane advantages (namely easy scalable mass production, no need of additional chemicals, energy efficient, and economical) and recent advancements make them more attractive. Major industries that employ membranes on a large scale are chemical, food, pharmaceutical, biotechnology, and tannery.

Membranes are classified into various categories based on their different traits, such as pore size, materials, synthesis, and cross-sectional attributes [3,5]. Based on pore size, membranes are classified as microfiltration, ultrafiltration, nanofiltration, or reverse osmosis. Materialistically, membranes can be organic or inorganic. Further, based on their method of synthesis, membranes can be classified into different types of membrane synthesis methods, such as phase inversion, dip-coating, sol-gel, track etching, etc. The membranes based on the configuration of their top and bottom layers, as can be seen from the cross-sectional attributes, are classified as symmetric, asymmetric, or mixed matrix.

Recently, much research was carried out in the development of effective and efficient stimuli-responsive membranes with controlled separation and permeation properties. As said earlier, the stimuli-responsive membranes show characteristic change in either their function or structure in response to an external stimuli, such as concentration, pH, temperature, pressure, light, electric, magnetic, ultrasound, etc. However, most of the external stimuli are practically ineffective. Therefore, only few of the selective external stimuli are suitable for membrane operations, such as pH, temperature, pressure, electric, light, and magnetic.

Furthermore, there are different types of responsive membranes based on their response to an external stimuli, for example pH responsive, temperature responsive, light responsive, electrical responsive, magnetic responsive, etc. [2]. These different types of stimuli-responsive membranes are effective in various types of applications due their specific responsiveness to an external stimuli. There are diverse needs and conditions in distinctive separation applications; therefore, these different types of stimuli-responsive membranes available provide us with a choice to select a membrane based on the requirement of a particular separation process.

1.2.1 pH-Responsive Membranes

The stimuli-responsive membranes that respond to an external pH stimuli are termed pH-responsive membranes [2]. These membranes show characteristic change in their function or structure on a change in pH. There are different pH-responsive groups, such as weak polyelectrolytes. The carboxyl and hydroxyl functional groups on change in pH exhibits a change in their confirmation by the virtue of protonation and deprotonation based on the value of the pH [6]. This mechanism will be discussed in the next section.

The inherit properties of a pH-responsive membrane make it a perfect choice for applications in food, pharma, and biotechnology sectors. The exploration of pH-responsive membranes for controlled drug delivery is one of the best examples. Similarly, a pH-based membrane separation process provides various options and is suitable for different tangible membrane-based separation processes.

1.2.2 Temperature-Responsive Membranes

The stimuli-responsive membranes that respond to an external temperature stimuli are known as temperature-responsive membranes [2,7]. These membranes behave differently at temperatures below and above the LCST (lower critical solution temperature) of the temperature-responsive group used for the synthesis of the stimuli-responsive membranes. For example, Singh et al. synthesized pH- and temperature-responsive membranes by using pH and thermo-responsive poly (N-vinylcaprolactam-TiO_2-acrylic acid) (VCL-TiO_2-AA) polymer nanocomposite [1]. The synthesized membranes showed a temperature-responsive behavior above and below 35°C, which is the LCST of N-vinylcaprolactam (VCL). The membrane water flux changes above and below the LCST of VCL; high at below LCST and low above LCST. Therefore, the use of VCL successfully made the membranes temperature responsive. The temperature-responsive membranes prove to be useful in the development of antifouling and self-cleaning membranes, as proposed by Singh et al. [1].

1.2.3 PHOTORESPONSIVE MEMBRANES

Photoresponsive membranes are responsive to external light stimulus [2]. There are numerous photoresponsive materials available that can be used to impart photo responsiveness to the membranes, such as materials with azobenzene [8]. These photoresponsive materials respond to light of specific intensity and wavelength and induce a functional or structural change. These changes are surface modification, swelling-shrinking, self-assembly and, the most important, the shape change. Further, the advantages associated with photoresponsive membranes make them more attractive, like photo responsiveness; photo responsiveness is safe for other constituents of the process because no additional chemicals are required for appropriate photo responsiveness, and it's economical. These properties make photoresponsive membranes user, process, and environment friendly. This is the reason stimuli-responsive membranes are so popular among the membrane enthusiasts.

1.2.4 BIOLOGICAL-RESPONSIVE MEMBRANES

Biomimicry is one of the art-inspired contributing effective solutions for day-to-day life. Method and techniques regularly develop taking inspiration from living organisms. It is both effective and efficient artform to develop new technology. Similarly, membranes are also developed to be biomimetic [2,3]. Further, biologically responsive membranes are synthesized using biomimetic materials deciphering the exact role played by the biological materials [9]. The development of analytical techniques, molecular biology, and supramolecular chemistry made it possible to develop biomimetic material and membranes [10]. In the case of biomimetic membrane development, the cell membrane is a great source of inspiration as it provides exceptional selective separation with high transport rates. These membranes are highly effective for the healthcare industry; for example artificial organs, dialysis, controlled drug delivery, etc.

1.2.5 ELECTRORESPONSIVE MEMBRANES

The membranes responsive to electric stimuli are called electroresponsive membranes [2]. These membranes observe changes in their functional and structural attributes in response to an electric stimulus. Electroactive materials, such as perfluorinated polymers and styrenic copolymers are used for the synthesis of electroresponsive membranes [11]. The electroresponsive membranes have shown potential in biomedical, semiconductor, and sensor development with improved antifouling property.

1.2.6 MAGNETIC FIELD RESPONSIVE MEMBRANES

Magnetic field responsive membranes respond to an external magnetic field [2]. The nanomagnetic materials and organometallic copolymers are used for the development of magnetic field responsive membranes. Again, these membranes are promising for applications in the field of biomedical, sensor, microelectronics, and smart textiles. Also, these membranes consist of improved antifouling properties due to the fact that they show a change in pore symmetry and membrane morphology under the influence of an external magnetic field [12].

1.2.7 ULTRASOUND-RESPONSIVE MEMBRANES

Membranes prepared using ultrasound-responsive smart materials and respond to an ultrasound stimulus are termed ultrasound-responsive membranes [2]. The membrane properties, such as pore size and porosity, can be changed by applications of ultrasounds as external stimuli. Further, ultrasound-responsive membranes show a great potential in the field of biomedical engineering. Ultrasound-induced cancer therapy is a new development in the field of sonochemistry, wherein these membranes could be very effective in terms of controlled drug delivery at a specific site [13].

This discussion on different types of stimuli-responsive membranes shows that these membranes are potentially suitable for various applications due to their reversible functional and structural properties. Further, the development of stimuli-responsive membranes diversify their scope and application. This increased the use of membranes on large-scale applications in various industries. Therefore, the development of stimuli-responsive membranes came as a boon for the field of membrane science.

1.3 RESPONSIVE MEMBRANE MATERIALS

The materials containing different functional groups that are responsive to various types of external stimuli are used to synthesize stimuli-responsive membranes. This section discusses various types of responsive materials used for the synthesis of different types of stimuli-responsive membranes; for example pH-, temperature-, photo-, electro-, magnetic-, and ultrasound-responsive membranes.

1.3.1 pH-RESPONSIVE MATERIALS

The pH-responsive materials induce pH-responsive behavior to the synthesized membranes. This section discusses the most commonly used pH-responsive materials for the synthesis of pH-responsive membranes. Generally, weak electrolytic polymers are good inducers of pH-responsive behavior; for example, polymers containing carboxyl (-COOH) and pyridine (C_5H_5N) [14–16]. The polymers with carboxyl groups swell in alkaline conditions because of carbonyl and hydroxyl functional groups. In alkaline conditions, carboxyl groups dissociate to form carboxylate ions that result in increased charge density. Further, this leads to swelling due to an increase in overall polymer volume. Consequently, under acidic conditions, protonation of the carboxyl groups takes place and leads to increased hydrophobic interactions. This results in shrinkage due to a decrease in overall polymer volume. However, in the case of pyridine-containing polymers, the results are opposite, that is, at higher and lower pH the polymer shrink and swell, respectively. Polyacrylic acid and polymethacrylic acid are the two most commonly used carboxyl-containing pH-responsive polymers. Further, polyvinyl pyridine is a common pyridine-containing pH-responsive polymer. Apart from carboxyl and pyridine pH-responsive groups, there are other examples, such as dibutyl amine, imidazole, and tertiary amine methacrylate [2]. The presence of these groups in membranes is responsible for change in their shape or configuration in response to a pH change. The pK_a values determine the response of the polymers under specific pH conditions; for example, polymers containing carboxyl and pyridine groups swell at a pH higher and lower than their pK_a values, respectively. Therefore, polymer pK_a values play an important role in factual use of the polymers for imparting appropriate pH-responsive behavior in the synthesized membranes. Thus, appropriate knowledge about the pK_a values of polymers is crucial for designing effective pH-responsive membranes. Further, it makes it easy to choose a polymer for desired pH-responsive membrane application.

1.3.2 TEMPERATURE-RESPONSIVE MATERIALS

The materials that respond to an external temperature change by change in their shape or conformation are known as temperature–responsive materials. These materials, when used to synthesize membranes, result in the formation of temperature responsive membranes. These materials contain a characteristic feature of having a lower critical solution temperature (LCST). For example, poly(N-isopropylacrylamide), commonly termed PNIPAAM, shows a characteristic LCST of 32°C [2]. Further, due to its thermo-responsive property, it is widely used for the synthesis of various temperature-responsive materials including membranes [1–3]. Additionally, poly-vinylcaprolactam (PVCL) is also a commonly used thermo-responsive polymer. The use of temperature-responsive materials with other polymers result in an increase or decrease in their LCST. The temperature-responsive materials are also used to synthesize dual- or multi-responsive materials by using materials that respond to a different external stimuli with the potential for different applications [1].

1.3.3 PHOTORESPONSIVE MATERIALS

The materials that respond to light of a specific wavelength are known as photo-responsive materials. These materials change their physicochemical characteristics, namely shape change, swelling-shrinkage, self-assembly, and surface modification by responding to light of a given wavelength and intensity. This property helps these materials to be used for diverse applications in different fields, such as healthcare, pharmaceutical, biotechnology, and environment.

The materials bearing azobenzene groups are considered the most common photoresponsive materials. These materials show photoisomerization that enables them to be photoresponsive in nature. The *cis* and *trans* forms of the compounds can be switched reversibly by using UV light of >400 nm and 300–400 nm, respectively, as shown in Figure 1.2.

Further, compounds containing azobenzene can be categorized into following types, namely azobenzene-type molecules, aminoazobenzene-type molecules, and pseudo-stilbenes [17]. The strong adsorption spectra of these molecules give them

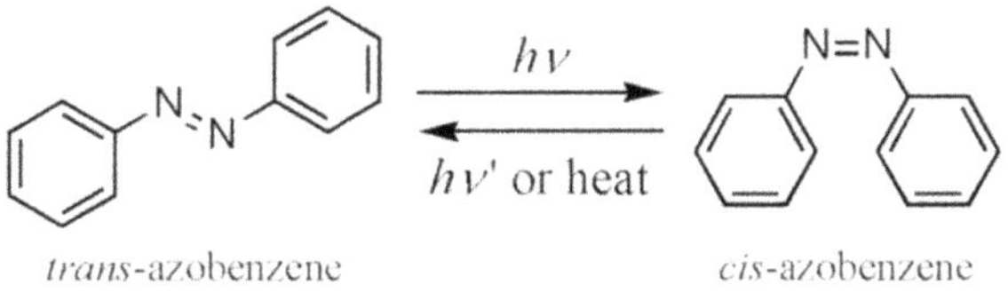

FIGURE 1.2   Schematic presentation of the photoisomerization of azobenzene.

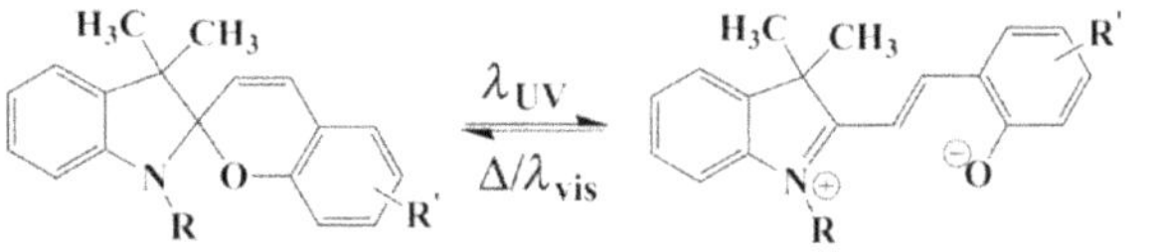

FIGURE 1.3   Photoresponsive behavior of spiropyran (reproduced with permission from Rim et al. [18] © American Chemical Society).

bright colors like yellow (azobenzene-type molecules), orange (aminoazobenzene-type molecules), and red (pseudo-stilbenes).

The *cis* and *trans* absorption spectra of azobenzene-type and aminobenzene-type molecules are different, that means, light of different wavelengths is required for their *cis* and *trans* and vice versa transformations. On the other hand, in the case of pseudo-stilbene molecules this is not the case, that is, a single wavelength of light can trigger *cis* and *trans* configurations of the molecules in forward (*trans* → *cis*) as well as reverse (*cis* → *trans*) direction. This makes pseudo-stilbene class of molecules exciting photoresponsive materials for various applications.

Additionally, the photoresponsive properties of the spiropyran-based materials were discovered in 1952 by Fischer and Hirshberg. Spiropyrans are organic compounds with two isomers (Figure 1.3) with novel features [18].

Further, spiropyrans-based materials can be used for diverse applications due to their response to different external stimuli, such as light, temperature, pH, mechanical, metal ions, and redox potential [2].

1.3.4   BIOLOGICAL-RESPONSIVE MATERIALS

Biological-responsive materials are biomimetic in nature that mimics the function or structure of naturally occurring biological materials [9]. Mostly, these designed and curtailed synthetically as per the need motivated by the properties and features of the natural material. The natural materials consist of efficient functions that are rare and unique. Therefore, there is a desire to synthesize such materials synthetically; for example, the lotus leaves, shark scales, and gecko sticking capability, etc.

The biological responsive materials can be categorized as:

i. Biomimetic hybrid: These materials consist of biological compounds responsible for functional properties of the materials.

ii. Biomimetic modified: These materials are synthetically synthesized and modified with functional molecules. This helps the material to mimic the desired function performed by the naturally occurring material.

iii. Biomimetic synthetic: These materials are completely developed synthetically, mimicking the functional and structural attributes of the natural materials.

These traits of the biological responsive materials make them a better choice for the development of various applications in different sectors with high efficiency and effectiveness.

1.3.5 ELECTRORESPONSIVE MATERIALS

Electroresponsive materials are also known as electro active, as they change their functional and structural attributes under the influence of an electric field. This property of electroresponsive materials made them perfect for various applications, namely sensors, actuators, optical systems, drug delivery, robotics, space, and energy [11].

The electroresponsive materials can be categorized under two categories: (i) ionic electroresponsive materials and (ii) dielectric electroresponsive materials [2]. In ionic electroresponsive materials, the electro responsiveness alters the local concentration of the free ions, in the presence of an electric field; in the material this exhibits the electroresponsive behavior of the material. Some of the examples of this category are conducting polymers, ionic polymers, gels, and ionic polymer metal composites. Further, in the dielectric electroresponsive materials, the coulombic (electrostatic) forces result in the deformation of the material that gives rise to the electro responsiveness in the material. Additionally, a contrasting feature among these electroresponsive materials is that the ionic electroresponsive materials show a low rate of reaction and small deformations and dielectric electroresponsive materials show a fast rate of reaction and high deformations. Further, the ionic electroresponsive materials can perform normally under wet conditions, whereas the dielectric electroresponsive materials operate under dry conditions.

Materials with high ion mobility and good mechanical properties make good electroresponsive materials; for example, perfluorinated polymers, styrenic copolymers, sulfonated polyimides. Further, non-ionic materials, such as poly(vinylidene fluoride), silicone, and polyacrylates, are also used as electroresponsive materials [19,20]. Additionally, bacterial cellulose is also explored as a biological alternative for electroresponsive materials [21]. This biopolymer consist of hydroxyl polar groups that make it an important electro active material. The use of such biopolymers further strengthens the sustainable growth of material science for various applications.

1.3.6 MAGNETIC FIELD RESPONSIVE MATERIALS

Magnetic field responsive materials are responsive materials that respond to an external magnetic field by changing their physicochemical characteristics [12]. These materials carry a wide scope in drug delivery, tissue engineering, sensors, textiles, fibers, actuators, and electromechanics. A recent example is artificial muscles developed by Kant et al. [22]. These muscles are synthesized by using liquid crystalline polymers and elastomers. The external magnetic as well as temperature-induced phase separation results in the movement of these artificial muscles in the form of volume change and deformation.

However, the available magnetic-responsive materials are not effective and efficient in terms of their use in an application. For example, the artificial muscles' response time is very high and this makes them unusable. Therefore, different methods are explored to enhance the effectiveness and efficiency of the magnetic-responsive materials. Doping of materials with magnetic nanoparticles is one of the simplest examples that is carried out to improve the overall magnetic responsive behavior of the materials. A few of the examples of magnetic nanoparticles are Fe_3O_4, Co, Ni, FeN, FePt, and FeP [2]. These nanoparticles generate small magnets in the material and thus enable them to respond to weaker stimuli.

1.3.7 ULTRASOUND-RESPONSIVE MATERIALS

Ultrasound-responsive materials respond to an external ultrasound field and show physicochemical changes [13]. The ultrasound-responsive materials show characteristic properties that can be used in various applications, such as biomedical, manufacturing, and others. The ultrasound came out as a great tool in biomedical imaging and navigation. This success makes it a great choice for non-invasive applications. Similarly, the ultrasound-responsive materials can be used for various medical procedures, for example, cancer therapy; herein, the ultrasound-responsive materials are introduced in the body and activated by the virtue of an external ultrasound field. The synergistic effect between the thermal and cavitation energies results in the detection or omission of cancerous cells. Purkait et al. discussed ultrasound-responsive materials in detail in their book [2].

1.4 SYNTHESIS METHODS

The membrane synthesis is an art that should be carried out with the outmost care. The perfectly followed membrane synthesis method will result in a membrane with perfect features and therefore, always sought after. There are different membrane synthesis methods for the fabrication of different membranes. Herein, mainly methods using coating and blending are discussed as they most closely represent the synthesis of stimuli-responsive membranes [2,3]. Further, the most prominent method for membrane synthesis by blending of membrane constituents, phase inversion, developed by Loeb and Sourirajan [23], is discussed.

1.4.1 COATING

Coating is a method that provides different options to synthesize membranes by either surface coating or grafting the membrane constituents over the membrane surface. The advantages of this method are that the coated materials will be available on the membrane surface and, therefore, provide better functionality compared to embedded

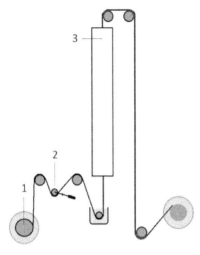

FIGURE 1.4 Schematic diagram of the continuous dip-coating device; 1 – hollow fiber spool, 2 – thread tension regulation, 3 – drying chimney (reproduced with permission from Jesswien et al. [25] © Elsevier).

materials [24,25]. This enhances the functionality of the synthesized membranes. Thus, the membrane provides better results in terms of its efficiency.

In the coating method of membrane synthesis, a thin layer of the membrane constituent is coated over a porous membrane support. The thickness of the coated layer ranges from 50–100 μm. This thickness reduces to only 0.5–2.0 μm after evaporation of the solvent. The important factor for the success of this method is the consistency of the coating solution and affinity between the porous support and the coating solution (membrane constituent). The scheme of the coating method is shown in Figure 1.4.

However, the major drawback of this method is the stability of the coated layer over the membrane surface [26]. This coated layer erodes from the membrane surface over time in a membrane separation process. Therefore, the life of the membrane is reduced and a need arises for the use of another method that can provide improved stability of the membrane constituents over a longer period of time. The answer for this problem is the blending method, wherein the membrane constituents are directly blended in the membrane casting solution. This method provides better stability and improved membrane life.

1.4.2 PHASE INVERSION

The phase inversion method is the most commonly used method for the fabrication of stimuli-responsive membranes. This method is based on the controlled separation of two phases. The concentrated phase solidifies immediately and the membrane is formed. The structure of the synthesized membrane depends upon the changes that have taken place in between the phase inversion and solidification. Figure 1.5 shows the systemic scheme of the phase inversion method. The phase inversion method is categorized into different categories based on the driving force, such as thermally induced phase separation, nonsolvent-induced phase separation, drying-induced phase separation, and vapor-induced phase separation.

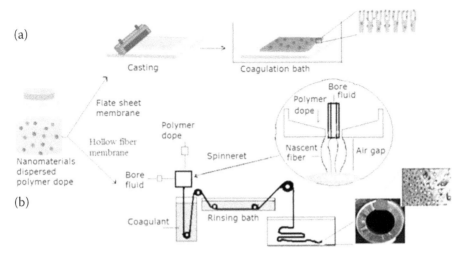

FIGURE 1.5 Schematic of (a) flat sheet and (b) hollow fiber MMM preparation by blending and phase inversion technique (reproduced with permission from Gohil and Choudhury [28] © Elsevier).

Phase inversion is the best suitable method for fabrication of membranes by blending of the membrane constituents [3,27]. The blending method supports the stable embeddedness of the materials into the membranes [26,28]. Further, the interactions and bonds among the membrane constituents help in the firm embeddedness of the materials in the membrane. Therefore, blending is the most common method to develop different types of membranes with greater stability and life span. Also, these membranes show improved efficiency and effectiveness for a longer period.

1.5 CHARACTERIZATION METHODS

The characterization of membranes is important as it determines their functionality and in turn suitability for a separation process. Membranes are characterized by using different standard characterization techniques in terms of their morphology, permeation, and functionality. The important characterization techniques for membranes are discussed in this section.

1.5.1 MORPHOLOGICAL ANALYSIS

The morphological analysis is an intelligent method to analyze the synthesized membranes. This method provides the accurate details of the membrane morphology that can be further used to assess the membranes for their accurate synthesis and production. Electron microscopy of different types, such as scanning electron microscopy, field emission electron microscopy, transmission electron microscopy, and atomic force microscopy, are widely used to assess the membrane morphology [2,3,5]. These techniques reveal the important technical details of the membranes, such as pore size, porosity, structure, and surface roughness of the membranes. The membranes

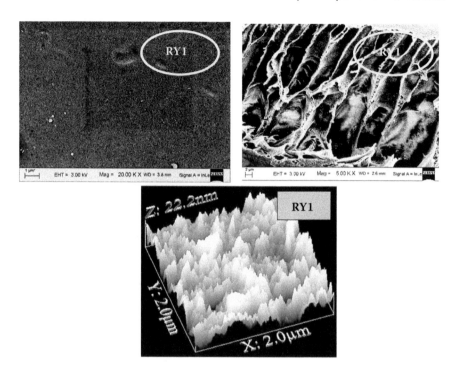

FIGURE 1.6 FESEM- and AFM-based morphological analysis of membranes (reproduced with permission from Singh et al. [16] © Elsevier).

were analyzed for their top surface as well as cross section to accumulate data to assess these membrane properties, as shown in Figure 1.6.

1.5.2 FUNCTIONAL ANALYSIS

Functional analysis is the analysis of membranes for the presence of functional characteristics. The membrane constituents used to modify membranes are detected for their presence in the membranes by using different techniques, such FTIR, XPS, TGA, Goniometer (Contact angle), XRD, EDX, etc. [3]. These techniques help to assess the membranes for the presence of these functional groups that attribute to their functionality. The early detection and confirmed presence of the functional groups affirm the correct synthesis of the membranes. This helps in the proper synthesis and use of the synthesized membranes for specific applications. Figure 1.7 shows membrane functional analysis with the FTIR technique.

1.5.3 PERMEATION-BASED ANALYSIS

The morphological and functional techniques provide details of the membrane surface and presence or absence of the functional groups. These techniques are not able to deduce any evidence about the inside of the membrane sample. Further, these techniques are not able to give accurate details about the membrane pores.

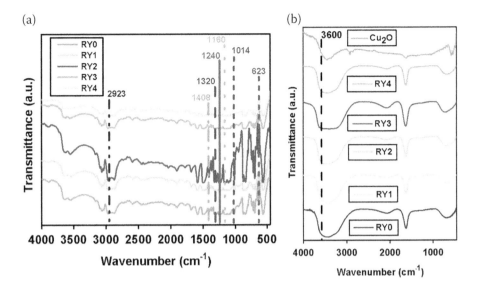

FIGURE 1.7 FTIR analysis of the membranes and stability analysis of Cu$_2$O photocatalyst in the membranes (reproduced with permission from Singh et al. [16] © Elsevier).

This is due to the fact that these techniques only analyze the membranes from the surface and not in depth. Therefore, the blind pores present in the membranes are not detected and, thus, lacks true pore details [3]. Hence, permeation-based techniques, such as pure water permeation, gas-liquid permeations, and liquid-liquid displacement porosimetry, are utilized to analyze the membranes. These techniques reveal the membrane flux, true pore details (pore size, pore number, and pore size distribution), hydraulic permeability, molecular weight cutoff, and separation efficiency [3]. Importantly, these techniques are non-destructive in nature and thus, the analyzed membrane can be further used.

The permeation-based techniques, to a great extent, provide accurate details about the membranes. However, the assumptions made during the fitting and interpretation of the membrane data in respective equations are not always true. Therefore, to an extent, the membrane values calculated by these techniques differ from the actual details. In retrospect, the calculated values provide a clear picture about the synthesized membranes. Furthermore, improved and developed techniques are required to assess the membranes with greater accuracy.

REFERENCES

[1] R. Singh, M. K. Sinha, and M. K. Purkait. Stimuli responsive mixed matrix polysulfone ultrafiltration membrane for humic acid and photocatalytic dye removal applications. *Separ. Purif. Technol.* 250 (2020) 117247.

[2] M. K. Purkait, M. K. Sinha, P. Mondal, and R. Singh. *Stimuli responsive polymeric membranes*, Elsevier, Academic Press, Cambridge, United States of America, 2018. ISBN: 9780128139615.

[3] M. K. Purkait and R. Singh. *Membrane technology in separation science*, CRC Press, Taylor & Francis, Boca Raton, 2018. ISBN: 1138626260, 9781138626263.

[4] S. Darvishmanesh, X. Qian, and S. R. Wickramasinghe. Responsive membranes for advanced separations. *Curr. Opin. Chem. Eng.* 8 (2015) 98–104.

[5] M. K. Purkait, R. Singh, P. Mondal, and D. Haldar. *Thermal induced membrane separation processes*, Elsevier, Academic Press, Cambridge, United States of America. ISBN: 9780128188019.

[6] M. K. Purkait, M. K. Sinha, P. Mondal, and R. Singh. pH-responsive membranes. In: *Stimuli responsive polymeric membranes*, Elsevier, Academic Press, Cambridge, United States of America, 2018. ISBN: 9780128139615, 10.1016/B978-0-12-813 961-5.00002-4.

[7] M. K. Purkait, M. K. Sinha, P. Mondal, and R. Singh. Temperature responsive membranes. In: *Stimuli responsive polymeric membranes*, Elsevier, Academic Press, Cambridge, United States of America, 2018. ISBN: 9780128139615, 10.101 6/B978-0-12-813961-5.00003-6.

[8] M. K. Purkait, M. K. Sinha, P. Mondal, and R. Singh. Photo responsive membranes. In: *Stimuli responsive polymeric membranes*, Elsevier, Academic Press, Cambridge, United States of America, 2018. 10.1016/B978-0-12-813961-5.00004-8.

[9] M. K. Purkait, M. K. Sinha, P. Mondal, and R. Singh. Biologically responsive membranes. In: *Stimuli responsive polymeric membranes*, Elsevier, Academic Press, Cambridge, United States of America, 2018. ISBN: 9780128139615, 10.101 6/B978-0-12-813961-5.00005-X.

[10] Y. -X. Shen, P. O. Saboe, I. T. Sines, M. Erbakan, and M. Kumar. Biomimetic membranes: a review. *J. Membr. Sci.* 454 (2014) 359–81.

[11] M. K. Purkait, M. K. Sinha, P. Mondal, and R. Singh. Electric field responsive membranes. In: *Stimuli responsive polymeric membranes*, Elsevier, Academic Press, Cambridge, United States of America, 2018. ISBN: 9780128139615, 10.101 6/B978-0-12-813961-5.00006-1.

[12] M. K. Purkait, M. K. Sinha, P. Mondal, and R. Singh. Magnetic responsive membranes. In: *Stimuli responsive polymeric membranes*, Elsevier, Academic Press, Cambridge, United States of America, 2018. ISBN: 9780128139615, 10.1016/ B978-0-12-813961-5.00007-3.

[13] M. K. Purkait, M. K. Sinha, P. Mondal, and R. Singh. Ultrasound responsive membranes. In: *Stimuli responsive polymeric membranes*, Elsevier, Academic Press, Cambridge, United States of America, 2018. ISBN: 9780128139615, 10.101 6/B978-0-12-813961-5.00008-5.

[14] M. K. Sinha and M. K. Purkait. Preparation and characterization of novel pegylated hydrophilic pH responsive polysulfone ultrafiltration membrane. *J. Membr. Sci.* 464 (2014) 20–32.

[15] Q. Wei, J. Li, B. Qian, B. Fang, and C. Zhao. Preparation, characterization and application of functional polyethersulfone membranes blended with poly (acrylic acid) gels. *J. Membr. Sci.* 337 (2009) 266–73.

[16] R. Singh, V. S. K. Yadav, and M. K. Purkait. Cu_2O photocatalyst modified anti-fouling polysulfone mixed matrix membrane for ultrafiltration of protein and visible light driven photocatalytic pharmaceutical removal. *Separ. Purif. Technol.* 212 (2019) 191–204.

[17] H. Rau. Photoisomerization of azobenzenes. In: J. Rebek (Ed.), *Photochemistry and photo-physics*, Boca Raton, FL: CRC Press, 1990.

[18] R. H. Rim, Y. Hirshberg, and E. Fischer. Mechanism of phototransformation in spiropyrans. *J. Phys. Chem. A* 66 (1962) 2470–7.

[19] A. Gugliuzza and E. Drioli. A review on membrane engineering for innovation in wearable fabrics and protective textiles. *J. Membr. Sci.* 446 (2013) 350–75.

[20] J. Lu, S. G. Kim, S. Lee, and I. K. Oh. Actuation of electroactive artificial muscle at ultralow frequency. *Macromol. Chem. Phys.* 212 (2011) 635–42.

[21] J. H. Jeon, I. K. Oh, C. D. Kee, and S. J. Kim. Bacterial cellulose actuator with electrically driven bending deformation in hydrated condition. *Sens. Actuators B: Chem.* 146 (2010) 307–13.

[22] P. G. de Gennes, M. Hébert, and R. Kant. Artificial muscles based on nematic gels. *Macromol. Symp.* 113 (1997) 39–49.

[23] S. Loeb and S. Sourirajan. Sea water demineralization by means of an osmotic membrane. *Adv. Chem. Ser.* 38 (1962) 117–32.

[24] R. Singh and M. K. Purkait. Evaluation of mPEG effect on the hydrophilicity and antifouling nature of the PVDF-co-HFP flat sheet polymeric membranes for humic acid removal. *J. Water Process Eng.* 14 (2016) 9–18.

[25] J. Jesswien, T. Hirth, and T. Schiestel. Continuous dip coating of PVDF hollow fiber membranes with PVA for humidification. *J. Membr. Sci.* 541 (2017) 281–90.

[26] R. Singh and M. K. Purkait. Role of poly(2-acrylamido-2-methyl-1-propanesulfonic acid) in the modification of polysulfone membranes for ultrafiltration. *J. Appl. Polym. Sci.* 134 (37) (2017) 45290.

[27] S. S. Ray, H. S. Bakshi, R. Dangayach, R. Singh, C. K. Deb, M. Ganesapillai, S. S. Chen, and M. K. Purkait. Recent developments in nanomaterials-modified membranes for improved membrane distillation performance. *Membranes* 10 (7) (2020) 140.

[28] J. M. Gohil and R. R. Choudhury. Chapter 2 – Introduction to nanostructured and nano-enhanced polymeric membranes: preparation, function, and application for water purification. In: *Nanoscale materials in water purification, a volume in micro and nano technologies*, Elsevier, Amsterdam, Netherlands, 2019, 25–57. 10.1016/B978-0-12-813926-4.00038-0.

2 pH-Responsive Membranes: Theoretical Concepts

2.1 INTRODUCTION TO pH-RESPONSIVE MEMBRANE THEORETICAL CONCEPTS

pH-responsive membranes are selective barriers for the separation of compounds under the influence of pH. The pH-based separation further enhances the selective separation of the pH-responsive membranes. This makes them popular for various applications in different fields, such as healthcare, food, and the environment. However, for proper utilization of these membranes, a better knowledge about their principles of separation operation and transport is required. Therefore, this chapter is dedicated to explaining the principle and transport-based concepts of pH-responsive membranes.

2.2 PRINCIPLE OF pH-RESPONSIVE MEMBRANE SEPARATION OPERATION

The membranes follow two models for their separation operation, namely solution-diffusion and pore flow (Figure 2.1) [1]. Generally, the solution-diffusion model (Fick's law) is followed by dense membranes with very minute pore sizes [2]. However, pH-responsive membranes follow the principle of the pore flow model for their separation operation. In this model, the separation of the feed compounds takes place through the membrane pores under the influence of the process driving force. The feed components pass in a convective flow across the membrane pores. Further, the size exclusion principle is followed for the separation of the feed components from each other; that is, the feed components with sizes less than the membrane pore size permeate across the membrane and vice versa.

The pressure gradient–driven convective flow is the basis of the pore flow model. Henry Philibert Gaspard Darcy (June 10, 1803–January 3, 1858), in 1856, formulated the famous Darcy's law that depicts this type of flow [3]. This law governs the feed flow across a membrane. Specifically, Darcy's law defines the pore flow model and is given by the following equation:

$$F_i = kc_i \frac{dp}{dx} \qquad 2.1$$

DOI: 10.1201/9781003201014-2

(a)   (b)

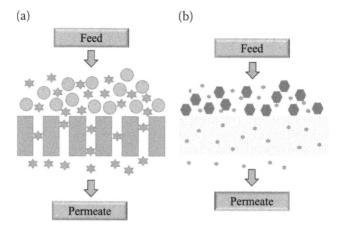

FIGURE 2.1 Membrane permeation models: (a) pore flow and (b) solution–diffusion.

wherein, dp/dx represents pressure gradient, c_i is the concentration of the i component, and k is the intrinsic permeability coefficient of membrane.

The pressure-driven membrane process with convective flow is a fast process of separation compared to solution diffusion. This is due to the fact that pores in membranes following the pore flow model are prominent, fixed, and interconnected. Further, an important relation between pore flow and membrane pore size is that the membrane pore size is proportional to the pore flow model characteristics. Importantly, membrane pores in the size range of 0.5–1.0 nm fluctuate between the solution diffusion or pore flow model.

2.2.1 KNUDSEN FLOW

The membrane pore size plays an important role in Knudsen flow as it is a pore size dependent flow. In this type of flow, the mean-free path of the feed components is equivalent to the membrane pore size that reduces the intra-collision of the feed components and results in better separation. The Knudsen flow takes place in membranes with a smaller pore size. Contrastingly, the membranes' larger pore sizes show viscous flow, wherein the intra-collision of the feed components is high, which results in ineffective separation. The two types of flows are shown in Figure 2.2.

The membrane flux for pH-responsive membranes under the Knudsen flow is given by the following relation:

$$F = \frac{\pi n r^2 D_k \Delta P}{RTl}$$

2.2

wherein R represents the gas constant, r is the membrane pore radius, D_k is the Knudsen diffusion coefficient, n is the molar concentration, and l is the membrane pore length. The Knudsen diffusion coefficient is equivalent to

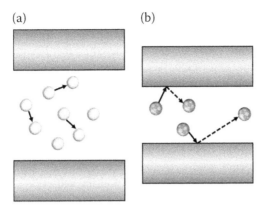

FIGURE 2.2   Schematic representation of (a) viscous flow and (b) Knudsen flow.

$$D_k = 0.66r\sqrt{\frac{8RT}{\pi M_w}}$$
   2.3

wherein r represents the membrane pore radius, T is the temperature, and M_w is the molecular weight. Eq. 2.3 shows the relation between the membrane flux and molecular weight. It defines the separation of two molecules and is inversely proportional to the ratio of the square root of the molecular weights.

2.2.2   VISCOUS FLOW

Viscous flow, as said previously, is not very efficient in terms of selective separation. This is due to the fact that the differential driving force available at the center or edges of the membrane pore is the main hydraulic separation force. Generally, viscous flow occurs in the case of a pressure drop across the membrane feed and permeates the sides. However, it also takes place when combined with some diffusion mechanisms that govern the basic separation process.

2.2.3   DRIVING FORCES IN pH-RESPONSIVE MEMBRANES

The main driving forces governing the separation in pH-responsive membranes are concentration, pressure, pH, and temperature. Further, these driving forces are related and governed by the chemical potential of the permeant. Therefore, the flux of pH-responsive membranes, F_i (kg/m^2 s) for a component i, can be given by the following relation:

$$F_i = -\emptyset_i \frac{d\mu_i}{dx}$$
   2.4

wherein $d\mu_i/dx$ represents the chemical potential of component i and $\emptyset_i$ is the coefficient of proportionality interlinking chemical potential and flux. Since, the pH-responsive membrane based separation processes are governed by more than one driving force, this relation (Eq. 2.4) is advantageous. Further, if only concentration and pressure gradients are considered as the driving forces in the pH-responsive membranes separation, then the chemical potential can be given by the following relation:

$$d\mu_i = RT \ln(\varepsilon_i \eta_i) + \sigma_i dp \qquad\qquad 2.5$$

wherein η_i represents the mole fraction of component i, ε_i is the coefficient of activity, p is the pressure, and σ_i is the molar volume of the component i. The coefficient of activity links the mole fraction to the activity.

The pressure plays no role in the change of volume of pH-responsive membranes; therefore, Eq. 2.5, by integrating with respect to concentration and pressure, becomes

$$\mu_i = \mu_i^{\circ} + RT \ln(\varepsilon_i \eta_i) + \sigma_i (p - p_i^{\circ}) \qquad\qquad 2.6$$

wherein μ_i° represents the chemical potential for the pure component i at pressure p_i°.

In membrane science, there are many assumptions made to define any permeation model. Mainly there are two commonly used assumptions: phases on feed and permeated sides are in equilibrium and the pressure applied is constant across the membrane system [1,4]. Therefore, as per these assumptions, it is only the concentration gradient that governs the chemical potential.

Considering these assumptions, Eqs. 2.4 and 2.5 are combined and, assuming σ_i as constant, gives the following relation:

$$F_i = -\frac{\varnothing_i \, RT}{\eta_i} \frac{d\eta_i}{dx} \qquad\qquad 2.7$$

The gradient of component i in Eq. 2.4 can be expressed in terms of mole fractions using the concentration term c_i (g/cm^3) that is defined as

$$C_i = M_{wi} \, \delta\eta_i \qquad\qquad 2.8$$

wherein M_{wi} represents the molecular weight of the component i (g/mol) and δ is the molar density (mol/cm^3). This gives to the following relation:

$$F_i = -\frac{\varnothing_i \, RT}{c_i} \frac{dc_i}{dx} \qquad\qquad 2.9$$

This relation resembles Fick's law and, therefore, can be written as

$$F_i = -D_i \frac{dc_i}{dx} \qquad\qquad 2.10$$

Finally, integrating Eq. 2.10 within the limits of membrane thickness (l) gives

$$F_i = \frac{D_i \left(c_{io(m)} - c_{il(m)} \right)}{l} \qquad 2.11$$

wherein $c_{io(m)}$ and c_{ilm} represent the concentration of the i component in the feed, which is in contact with the membrane at the feed and permeated interface, respectively.

2.3   MASS TRANSPORT IN pH-RESPONSIVE MEMBRANES

The pressure-driven convective flow is used to describe mass transport in pH-responsive membranes based on the pore flow model. Herein, the membrane selectivity depends on the membrane pore size for effective separation. The membranes follow the sieving process for separation of the feed components. Briefly, the feed components of a size larger than the membrane pore size are rejected and the feed components with a small size are allowed to permeate across the membrane.

There are two approaches to describe the separation and permeability characteristics of pH-responsive membranes. In the first approach, the feed components comprise of near spherical particles and the Carmen-Kozney equation [5] is applied as given in the following relation:

$$F = \frac{\varepsilon^3}{K\mu S^2 (1 - \varepsilon)^2} \frac{\Delta P}{l_{pore}} \qquad 2.12$$

wherein F represents the membrane flux, ε is the surface porosity, μ is the dynamic permeate viscosity, S is the specific area per unit volume, ΔP is the pressure drop across the membrane, l_{pore} is the membrane thickness, and K is the permeability constant. K and S both are membrane porosity dependent. Further, K is a function of membrane structure as well as other characteristics, namely membrane porosity, pore size distribution, and permeate viscosity.

In the second approach, the laminar flow is assumed across the membranes and differential momentum balance equation, known as Newton's law of viscosity, and can be used to describe it. Finally, after integration, the Hagen-Poiseuille equation [6] for average convective velocity across the membrane is obtained and can be used to calculate the membrane flux as given in the following relation:

$$F = \frac{\varepsilon^3 d_{pore}^2}{32 \, \mu \, \tau} \frac{\Delta P}{l_{pore}} \qquad 2.13$$

wherein τ represents the membrane tortuosity. The flux is again inversely proportional to the permeated viscosity. Further, as discussed earlier, Darcy's law can be applied to calculate the flow rate across the membrane. Additionally, it helps to calculate the membrane area required for targeted separation under given conditions, as shown in the following relation:

$$Q = \frac{-kA}{\mu} \frac{(P_b - P_a)}{L}$$ 2.14

wherein Q represents the total permeated flow (m³/s), k is the empirical constant for membrane permeability (m²), A is the membrane cross-sectional area to the flow, $(P_b - P_a)$ is the pressure drop across the membrane, μ is the viscosity (kg/ms), and L is the membrane thickness (m).

2.4 EFFECT OF pH ON MEMBRANE SEPARATIONS

The pH plays a vital role in the membrane-based separation processes. The pH affects the membrane surface charge, hydrophilicity or hydrophobicity, rejection, and efficiency. Therefore, the pH-responsive membranes' effectiveness and efficiency are widely governed by the pH of the membrane system.

2.4.1 SURFACE CHARGE

The membrane surface, due to the material composition, could be positively or negatively based on the pH value, as shown in Figure 2.3 [7,8]. For example, a membrane containing dissociable carboxylic and amine groups shows negative and positive charge under alkaline and acidic pH conditions, respectively [9–11].

Further, it is important that the functional groups be present on the membrane surface. Otherwise, the pH effect will not be prominent and that may affect the pH-based separation applications of the membranes. For example, a sulphonated polyethersulphone membrane containing dissociable sulphonic groups needs to be highly

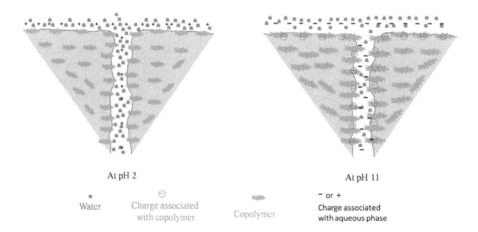

FIGURE 2.3 Schematic presentation for the mechanism of pH-responsive phenomenon of membranes (reproduced with permission from Sinha and Purkait [8] © Elsevier).

negatively charged. However, in actuality, this is not the case as the membranes show less negative charge at a neutral pH in comparison to membranes containing dissociable carboxylic and amine groups. This is due to the fact that either the sulphonic groups are not present completely over the membrane surface but embedded in the membrane matrix, or the sulphonation process is incomplete and that results in a very low density of sulphonic groups over the membrane surface [11]. Therefore, not only the pH but the presence of the dissociable groups in the membrane are important to show pH responsiveness [12]. Hence, pH-responsive membrane needs to be fabricated accurately to obtain greater efficiency. Further, by this knowledge, membranes with tailored pH responsiveness can be synthesized.

2.4.2 Hydrophilicity/Hydrophobicity

The hydrophilicity or hydrophobicity play an important role in the separation efficiency of the membranes. This surface property is also affected by the change in pH of the membrane system. This is because the dissociable groups vary the interaction between the membrane surface and water molecules that change the hydrophilicity or hydrophobicity of the membranes. Generally, the degree of dissociation of these groups increases with an increase in the pH value. Mostly the hydrophilicity of the membranes change with a change in the pH with commonly used dissociable groups, such as the carboxylic group. However, it totally depends upon the type and place where the dissociable group is present in the membranes. For example, the polyethersulphone (PES) membrane with dissociable sulphonic groups must show hydrophilic characteristics as a function of pH [11]. This is due to the addition of sulphonic groups to the hydrophobic PES membrane must improve its hydrophilic character. However, the results were opposite; that is, the membrane showed a hydrophobic character as a function of pH. This is due to the fact that the dissociable sulphonic groups were not completely present on the membrane surface or incomplete sulphonation. Therefore, the membranes' hydrophilic or hydrophobic character is also affected as a function of pH. This attribute can be used greatly in a pH-responsive membrane for developing membranes for diverse applications.

2.4.3 Membrane Permeability and Separation Efficiency

The permeability and separation efficiency of membranes is definitely affected by the pH of the membrane system, as shown in Figure 2.4 [13]. This is due to the fact that in the case of pH-responsive membranes, the membrane pore size differs with variation in the pH value [13,14].

The membrane pore size depends upon the type of dissociable groups present in the membrane that governs the surface charge of the membranes, and either increases or decreases. This phenomenon of pH-dependent membrane pore size change occurs due to the shrinking or relaxing of the pH-responsive groups present in the membrane. In the case of shrinking, the membrane pore size increases, which leads to increased permeability and reduced separation efficiency. On the other hand, in the case of relaxing, the membrane pore size reduces, which results in decreased membrane permeability but improved separation efficiency. These

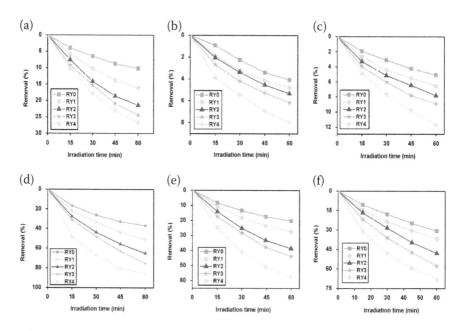

FIGURE 2.4 Photocatalytic removal of IBP under different conditions, dark: (a) acidic, (b) basic, (c) neutral; and simulated solar light: (d) acidic, (e) basic, (f) neutral (Purkait et al. 2018 [12] © Elsevier).

changes in membrane permeability and separation efficiency occur with a change in membrane pore size due to the fact that in case of larger membrane pore size the membrane flux is high and that gives high permeability and reduces the efficiency to retain feed components. Conversely, the small pore size reduces membrane flux and results in lower membrane permeability but enhances the membrane efficiency to retain feed components that are responsible for improved membrane separation efficiency. Further, the charge present on the feed components also affects the separation efficiency of the membranes, since a feed component will either be repelled or attracted based on the charge it carries with respect to the membrane. For example, a positively charged feed component will be attracted by the negatively charged membrane, and vice versa.

2.4.4 Reversibility of pH Effect

The pH effect is almost reversible; that is, on providing the original pH conditions, the membrane behaves accordingly. For example, if the membrane property is to give average permeability and separation efficiency at a neutral pH, and is used under alkaline conditions giving high permeability and low separation efficiency, then if it is again used at neutral pH, then it will provide the same results, which are average permeability and separation efficiency. Therefore, the pH effect is reversible and the membranes can be restored completely to their respective operating conditions.

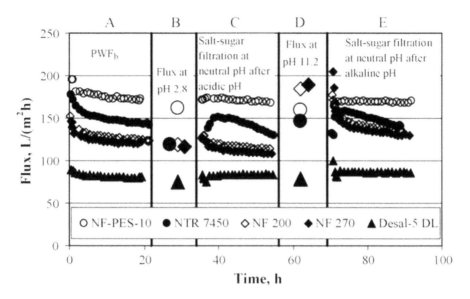

FIGURE 2.5 Fluxes of some NF membranes studied during different filtration stages. T = 40°C, Δp = 8 bar, 225 mg/L NaCl and 225 mg/L glucose. (A) Pure water flux, (B) flux of the salt–sugar solution at pH 2.8 when pH has been gradually shifted down from neutral, (C) flux of the salt–sugar solution at neutral pH, (D) flux of the salt–sugar solution at pH 11.2 when pH has been gradually shifted up from neutral, (E) flux of the salt–sugar solution at neutral pH. Cross-flow velocity 4 m/s (reproduced with permission from Manttari et al. [11] © Elsevier).

The reversibility of the membrane flux is presented by Manttari et al. in their pH-dependent study and shown in Figure 2.5 [11].

In this study, the fresh membrane samples were pressurized for long periods so that the flux could be stabilized. Later, a sugar–salt solution was permeated across the membrane while decreasing the pH up to 2.8. Thereafter, the sugar–salt solution was permeated across the membrane at a neutral pH for a long period of time. Later on, the pH of the sugar–salt solution became alkaline up to 11.2 gradually and the flux was measured. Thereafter, again the membranes were permeated at a neutral pH for a long duration and measured the flux. This study reveals that stable fluxes were obtained even after operating the membranes under variable pH conditions for longer periods, which proves the reversibility of the pH effect over membranes.

REFERENCES

[1] M. K. Purkait and R. Singh. *Membrane technology in separation science*, CRC Press, Taylor & Francis, Boca Raton, 2018. ISBN: 1138626260, 9781138626263.

[2] A. Fick. Über diffusion. *Poggendorff's Ann. Phys. Chem.* 94 (1855) 59.

[3] H. Darcy. *Les fontaines publiques de la ville de Dijon*, Dalmont, Paris, 1856.

[4] M. Mulder. *Basic principles of membrane technology*, Springer, Amsterdam, Netherlands, 2007.

[5] J. Kozeny. Ueber kapillare Leitung des Wassers im Boden. *Sitzungsber. Kais. Akad. Wiss., Wien* 136 (2a) (1927) 271–306.

[6] S. P. Sutera and R. Skalak. The history of Poiseuille's law. *Annu. Rev. Fluid Mech.* 25 (1993) 1–19.

[7] M. K. Purkait, M. K. Sinha, P. Mondal, and R. Singh. *Stimuli responsive polymeric membranes*, Elsevier, Academic Press, Cambridge, United States of America, 2018. ISBN: 9780128139615.

[8] M. K. Sinha and M. K. Purkait. Preparation and characterization of novel pegylated hydrophilic pH-responsive polysulfone ultrafiltration membrane. *J. Membr. Sci.* 464 (2014) 20–32.

[9] V. Freger, J. Gilron, and S. Belfer. TFC polyamide membranes modified by grafting of hydrophilic polymers: an FT-IR/AFM/TEM study. *J. Membr. Sci.* 209 (2002) 283–92.

[10] V. Freger, A. Pihlajamáki, Y. Shabtai, and J. Gilron. Distribution of fixed charge functional groups in the polyamide composite membranes. In: *Supplementary book of abstracts*, ICOM'02, Toulouse, France, 2002, p. 144.

[11] M. Manttari, A. Pihlajamaki, and M. Nystrom. Effect of pH on hydrophilicity and charge and their effect on the filtration efficiency of NF membranes at different pH. *J. Membr. Sci.* 280 (2006) 311–20.

[12] M. K. Purkait, M. K. Sinha, P. Mondal, and R. Singh. pH-responsive membranes. In: *Stimuli responsive polymeric membranes*, Elsevier, Academic Press, Cambridge, United States of America, 2018. ISBN: 9780128139615, 10.1016/B978-0-12-813 961-5.00002-4.

[13] R. Singh, V. S. K. Yadav, and M. K. Purkait. Cu_2O photocatalyst modified antifouling polysulfone mixed matrix membrane for ultrafiltration of protein and visible light driven photocatalytic pharmaceutical removal. *Separ. Purif. Technol.* 212 (2019) 191–204.

[14] R. Singh, M. K. Sinha, and M. K. Purkait. Stimuli responsive mixed matrix polysulfone ultrafiltration membrane for humic acid and photocatalytic dye removal applications. *Separ. Purif. Technol.* 250 (2020) 117247.

3 pH-Responsive Membrane Types, Modes, and Methods of Modifications

3.1 INTRODUCTION

Presently, there is a lot of research and development of pH-responsive membranes. This is due to the characteristic properties of the pH-responsive membranes, mainly the control over the membrane permeability and separation efficiency. These properties make them suitable for various industrial applications. The healthcare, biotechnology, food, and environment industries are the hubs for pH-responsive membrane applications. In this chapter, different types of pH-responsive membranes along with their mode of operations and modification methods are discussed in detail. This chapter will help better understand the field of pH-responsive membranes.

3.2 FLAT pH-RESPONSIVE MEMBRANES

Flat-sheet pH-responsive membranes are most commonly synthesized and used membranes. These membrane types are widely used on a lab scale for testing and validation purposes [1]. The membranes that are prepared in this configuration play an important role in the research and development of membrane science. The pH-responsive membranes are synthesized in flat-sheet configuration by either blending or surface coating the pH-responsive groups in the membranes. Further, for the membranes to show pH responsiveness, these pH-responsive groups must be present on the membrane surface. Otherwise, the membranes may not show pH-responsive characteristics effectively and efficiently.

3.3 HOLLOW FIBER pH-RESPONSIVE MEMBRANES

The hollow fiber pH-responsive membranes show immense potential for commercial and industrial applications. This is due to the reason that these membranes provide a larger surface area to volume ratio, which results in high packing densities [1–3]. Also, they possess less fouling and provide a longer life span. Further, they are advantageous to use in cross-flow mode as compared to flat-sheet membranes. These advantages are associated with hollow fiber membranes that make them highly useful for various membrane-based separation applications. The membrane science gained a big response from the industrial sector due to the development of the hollow fiber membranes.

DOI: 10.1201/9781003201014-3

3.4 MODE OF OPERATION

The membrane separation processes can be operated in different modes, namely batch and continuous. These modes of operation are discussed in this section.

3.4.1 BATCH

Batch operation is the most commonly used membrane separation operation and shown in Figure 3.1(a) [1]. The batch mode of a membrane operation is carried out with limited feed supply and, therefore, suitable for lab or pilot scale applications. The feed runs from a membrane cell of fixed quantity and thus, the membrane operation runs until the feed runs out. This mode of operation is analogous to a batch reactor. Further, this mode of operation gives time for membrane cleaning and maintenance of the membrane cell in between the two runs. This results in enhanced membrane life and efficiency for longer periods of time.

3.4.2 CONTINUOUS

The continuous mode of operation is a suitable method for industries as in this method the feed and permeate are continuously added and removed, respectively. In this method, the volumes to be handled are high and therefore suited for industrial applications. However, the continuous and long time of operation increases the membrane fouling and decreases the overall efficiency of the system over time. The continuous mode of operation is shown in Figure 3.1(b).

3.5 *IN-SITU* MODIFICATIONS

In-situ modifications mean modifications carried out inside the membranes; for example, blending of different pH-responsive groups inside the membranes by mixing them in the membrane casting solution. These methods provide stable and efficient binding of the additives in the membranes [4,5]. Further, this stable binding of the additives inside of the membranes enhances their functionality and stability. This results in their reuse for longer periods of time. In this section, the blending method is discussed in detail.

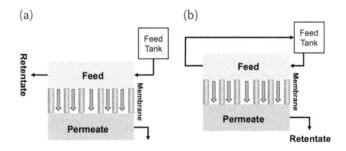

FIGURE 3.1   Schematic representation of (a) batch and (b) continuous membrane separation process.

3.5.1 Blending Method

The blending method is the best suitable method for membrane synthesis. This method allows synthesized membranes to be uniformly distributed and firmly embedded additives [6]. Further, the method possess less risk of additive leakage from the membranes during a membrane operation [7]. These are the reasons that makes the blending method a widely used membrane synthesis method.

Singh et al. used the blending method for the fabrication of stimuli-responsive membranes by adding pH and thermo-responsive poly (N-vinylcaprolactam-TiO$_2$-acrylic acid) (VCL-TiO$_2$-AA) polymer nanocomposite to the membranes [6]. The synthesized membranes show photocatalytic, pH, and temperature responsive properties. The study also confirmed that the additive nanocomposite is firmly embedded in the membrane matrix. Further, Singh et al. used a Cu$_2$O photocatalyst for the modification of polysulfone membranes for protein and pharmaceutical removal applications [7]. The study shows a stable presence of the additive in the membrane matrix with photocatalytic properties. Sinha and Purkait also developed pH-responsive polysulfone ultrafiltration membranes by using the novel cross-linked pegylated functional copolymer poly(acrylic acid-co-polyethylene glycol methyl ether methacrylate) [8]. The membranes show pH-dependent membrane permeability and separation properties. Wie et al. used poly(acrylic acid) gel to synthesize pH-responsive membranes [9]. The synthesized membranes successfully show pH-responsive behavior with a pH valve effect between pH 3 and 8. The pH-responsive behavior of the membranes was confirmed by assessing the membrane flux at different pH values. The membrane flux was measured as 138 mL/hm^2 mmHg at pH 2.3 and 40 mL/hm^2 mmHg at pH 8.6. Further, this membrane flux behavior is due to the additive (poly(acrylic acid) functionality as its pK$_a$ value lies between 4.3 and 4.9. Therefore, at a pH lower than the pK$_a$ value of the additive, the membranes' pore size increases due to the presence of carboxyl groups in their unionized state. On the other hand, in a case of a pH higher than the pK$_a$ value of the additive, the membrane pore size decreases due to the dissociation and extension of carboxyl groups. Lastly, the membranes were studied for their ion exchange properties with Cu^{2+} ions. The pH-responsive membranes show a 99.6% Cu^{2+} ion adsorption. This confirms that the membrane can successfully be used as an ion-exchange membrane. Additionally, Qian et al. [10] and Zou et al. [11] synthesized pH-responsive hollow fiber membranes using the blending method. The membranes successfully show a reversible pH-responsive behavior with an improved membrane permeability and separation profiles.

These studies confirm the effectiveness of the blending method for the synthesis of different types of membranes based on the use of various additives.

3.6 *EX-SITU* MODIFICATIONS

The ex-situ modifications are the modifications carried out at the outside (surface) of the membranes. These modifications are namely surface coating, surface grafting, and membrane pore filling [1,2]. These methods are promising in the modification of membranes for various applications. The advantages of these methods are that the additive used to modify the membranes remains on the membrane surface that helps to exert the selective functionality appropriately in the membrane separation process. In this section, these membrane surface modification methods are discussed individually in detail for their better understanding.

3.6.1 SURFACE COATING

The surface coating method is used to modify membranes with additives coated over the membrane surface. This method is advantageous due to its effective control and high frequency of coating additives onto the membrane surface [12,13]. Thus, it is an appropriate method to be used for the synthesis of pH-responsive membranes by coating pH-responsive additives over the membrane surface.

This method is successfully used to synthesize pH-responsive drug delivery membranes [14]. The surface coating of poly(2-hydroxyethyl methacrylate) membranes was carried out by using acrylic acid and 2-(diethylamino)ethyl methacrylate. The synthesized membranes were analyzed for the delivery of model drugs with different molecular weights and charges. The study confirms successful delivery of the drugs. Further, this study states that the interaction between the drug and membrane functional groups also governs the drug delivery process. Wang et al. successfully synthesized pH-responsive polyacrylonitrile hollow fiber membranes by using surface coating method [15]. Further, different polymers were used to synthesize pH-responsive membranes in different configurations, such as a flat sheet and hollow fiber using the coating method; for example, poly(2-vinylpyridine) [16], poly(allylamine hydrochloride), and poly(sodium 4-styrenesulfonate) [17]. The method is successful in imparting reversible pH responsiveness to the membranes with a great success rate. However, the only drawback of this method is that the additives' surface coated over the membrane surface show lower stability as they come off over time during the membrane process. Therefore, there is a need to develop methods and techniques to enhance the stability of the coated additives over the membrane surface. This will make it the most suitable membrane modification method with the availability of numerous variations (chemical reaction methodologies) to carry out membrane modifications.

3.6.2 SURFACE GRAFTING

The surface grafting method is used to graft pH-responsive groups to the membrane or membrane pore surface [2,18]. There are various methods to carry out grafting; namely, redox grafting, Atom Transfer Radical Polymerization (ATRP), Reversible Addition-Fragmentation chain-Transfer polymerization (RAFT), etc. The grafting method can be "grafting to" or "grafting from" [19,20]. In the "grafting to" method, direct grafting of the pH-responsive group over the membrane surface takes place. On the other hand, in "grafting from" the grafting reaction starts from the pH-responsive group and results in the grafting of themselves over the membrane surface. The important components of a grafting system are monomer, initiator, catalyst, and solvent. These components carry out the grafting reaction and modify the membranes. The surface grafting is carried out by introducing active sites over the membrane surface induced by free radicals, temperature, light, ozone, or plasma. Later, monomer polymerization is initiated and carried out by using an initiator and a catalyst, respectively [21].

3.6.3 Membrane Pore Filling

The membrane pore filling method results in the accumulation of the pH-responsive additives or groups on the membrane pore surface, as shown in Figure 3.2. This leads to the showcase of a brilliant pH-dependent valve effect and efficient separation by the membranes. This method of membrane modification is prominent as the membrane permeability as well as separation property depends upon the membrane pores [22]. Therefore, adding pH-responsive groups over the membrane pores give them additional capability to carry out the membrane functions effectively. The research carried out with this method confirms that this position for adding pH-responsive groups is the best for making the most of the pH-responsive membrane property for various applications.

The polyethylene and polypropylene microfiltration membrane pores were filled with a polyelectrolyte to impart pH responsiveness to the membranes [23]. The synthesized membranes show excellent pH-responsive behavior. The microfiltration membranes with a simple change in pH converts reversibly to reverse osmosis membrane. This showcases the ability of the pore filling method to be more effective than the blending and surface coating methods for the modification of the membranes. In another study, reverse osmosis membranes with improved performance were synthesized by cross-linking methyl acrylate using the pore filling method [24]. The additive methyl acrylate was filled in the membrane pores by grafting it over the high-density polyethylene membrane surface. The synthesized membranes show better stabilization and performance at high pressures up to 12

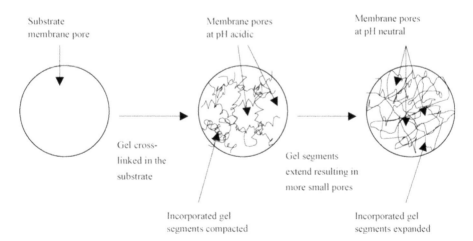

FIGURE 3.2 Schematic representation of the relative pore size and pore density between the nascent membrane and the pore-filled membrane swollen gel at pH acidic and pH neutral (reproduced with permission from Hu and Dickson [22] © Elsevier).

MPa for the separation of chloroform and n-hexane. Further, microfiltration membranes were explored for the operation of nanofiltration membranes by the pore filling method [25]. The polypropylene microfiltration membranes were filled with different additives; namely poly(4-vinylpyridine), poly(vinylbenzylchloride), or acrylic acid. These modified membranes were then analyzed for their various characteristic parameters, such as modification, permeability, and separation properties. This study shows that the pore filling method is effective and efficient in imparting pH responsiveness and other characteristic features to the membranes. Further, these studies confirm that the pore filling method allows the membranes to be operated with variable pore sizes; for example, the microfiltration membrane could be a nanofiltration or reverse osmosis membrane. Hence, these membranes can be used for various applications, such as separation of salts, organic mixtures, metal ions, and drug delivery.

REFERENCES

[1] M. K. Purkait and R. Singh. *Membrane technology in separation science*, CRC Press, Taylor & Francis, Boca Raton, 2018. ISBN: 1138626260, 9781138626263.

[2] M. K. Purkait, M. K. Sinha, P. Mondal, and R. Singh. *Stimuli responsive polymeric membranes*, Elsevier, Academic Press, Cambridge, United States of America, 2018. ISBN: 9780128139615.

[3] M. K. Purkait, R. Singh, P. Mondal, and D. Haldar. *Thermal induced membrane separation processes*, Elsevier, Academic Press, Cambridge, United States of America, ISBN: 9780128188019.

[4] R. Singh and M. K. Purkait. Evaluation of mPEG effect on the hydrophilicity and antifouling nature of the PVDF-co-HFP flat sheet polymeric membranes for humic acid removal. *J. Water Process Eng.* 14 (2016) 9–18.

[5] R. Singh and M. K. Purkait. Role of poly(2-acrylamido-2-methyl-1-propanesulfonic acid) in the modification of polysulfone membranes for ultrafiltration. *J. Appl. Polym. Sci.* 134 (37) (2017) 45290.

[6] R. Singh, M. K. Sinha, and M. K. Purkait. Stimuli responsive mixed matrix polysulfone ultrafiltration membrane for humic acid and photocatalytic dye removal applications. *Separ. Purif. Technol.* 250 (2020) 117247.

[7] R. Singh, V. S. K. Yadav, and M. K. Purkait. Cu$_2$O photocatalyst modified antifouling polysulfone mixed matrix membrane for ultrafiltration of protein and visible light driven photocatalytic pharmaceutical removal. *Separ. Purif. Technol.* 212 (2019) 191–204.

[8] M. K. Sinha and M. K. Purkait. Preparation and characterization of novel pegylated hydrophilic pH-responsive polysulfone ultrafiltration membrane. *J. Membr. Sci.* 464 (2014) 20–32.

[9] Q. Wei, J. Li, B. Qian, B. Fang, and C. Zhao. Preparation, characterization and application of functional polyethersulfone membranes blended with poly (acrylic acid) gels. *J. Membr. Sci.* 337 (2009) 266–73.

[10] B. S. Qian, J. Li, Q. Wei, P. L. Bai, B. H. Fang, and C. S. Zhao. Preparation and characterization of pH-sensitive polyethersulfone hollow fiber membrane for flux control. *J. Membr. Sci.* 344 (2009) 297–303.

[11] W. Zou, Y. Huang, J. Luo, J. Liu, and C. S. Zhao. Poly (methyl methacrylate–acrylic acid–vinylpyrrolidone) terpolymer modified polyethersulfone hollow fiber membrane with pH-sensitivity and protein antifouling property. *J. Membr. Sci.* 358 (2010) 76–84.

[12] Q. N. T. Xuan, S. S. Chen, H. M. Chang, T. Cao, and R. Singh. Effects of polyethylene glycol and glutaraldehyde cross-linker on TFC-FO membrane performance. *Environ. Technol. Innov.* 20 (2020) 101059.

[13] S. S. Ray, H. S. Bakshi, R. Dangayach, R. Singh, C. K. Deb, M. Ganesapillai, S. S. Chen, and M. K. Purkait. Recent developments in nanomaterials-modified membranes for improved membrane distillation performance. *Membranes* 10 (7) (2020) 140.

[14] L.-T. Ng and K.-S. Ng. Photo-cured pH-responsive polyampholyte-coated membranes for controlled release of drugs with different molecular weights and charges. *Radiat. Phys. Chem.* 77 (2008) 192–9.

[15] N. Wang, G. Zhang, S. Ji, Z. Qin, and Z. Liu. The salt-, pH- and oxidant-responsive pervaporation behaviors of weak polyelectrolyte multilayer membranes. *J. Membr. Sci.* 354 (2010) 14–22.

[16] M. Orlov, I. Tokarev, A. Scholl, A. Doran, and S. Minko. pH-responsive thin film membranes from poly(2-vinylpyridine): water vapor induced formation of a microporous structure. *Macromolecules* 40 (2007) 2086–91.

[17] D. Lee, A. J. Nolte, A. L. Kunz, M. F. Rubner, and R. E. Cohen. pH-induced hysteretic gating of track-etched polycarbonate membranes: swelling/deswelling behavior of polyelectrolyte multi-layers in confined geometry. *J. Am. Chem. Soc.* 128 (2006) 8521–9.

[18] M. K. Purkait, M. K. Sinha, P. Mondal, and R. Singh. pH-responsive membranes. In: *Stimuli responsive polymeric membranes*, Elsevier, Academic Press, Cambridge, United States of America, 2018. ISBN: 9780128139615, 10.1016/B978-0-12-813 961-5.00002-4.

[19] Y. F. Duanna, Y. C. Chena, J. T. Shena, and Y. H. Lin. Thermal induced graft polymerization using peroxide onto polypropylene fiber. *Polymer* 45 (2004) 6839–43.

[20] W. Wang, L. Wang, X. Chen, Q. Yang, T. Sun, and J. Zhou. Study on the graft reaction of poly(propylene) fiber with acrylic acid. *Macromol. Mater. Eng.* 291 (2006) 173–80.

[21] Z. Xu, J. Wang, L. Shen, D. Men, and Y. Xu. Microporous polypropylene hollow fiber membrane part I. Surface modification by the graft polymerization of acrylic acid. *J. Membr. Sci.* 196 (2002) 221–9.

[22] K. Hu and J. M. Dickson. Modelling of the pore structure variation with pH for pore-filled pH-sensitive poly(vinylidene fluoride)–poly(acrylic acid) membranes. *J. Membr. Sci.* 321 (2008) 162–71.

[23] A. M. Mika, R. F. Childs, J. M. Dickson, B. E. McCarry, and D. R. Gagnon. A new class of polyelectrolyte-filled microfiltration membranes with environmentally controlled porosity. *J. Membr. Sci.* 108 (1995) 37–56.

[24] T. Kai, H. Goto, Y. Shimizu, T. Yamaguchi, S. Nakao, and S. Kimura. Development of crosslinked plasma-graft filling polymer membranes for the reverse osmosis of organic liquid mixtures. *J. Membr. Sci.* 265 (2005) 101–7.

[25] R. F. Childs, A. M. Mika, A. K. Pandey, C. McCrory, S. Mouton, J. M. Dickson. Nanofiltration using pore-filled membranes: effect of polyelectrolyte composition on performance. *Separ. Purif. Technol.* 22–23 (2001) 507–17.

4 Materials and Synthesis of pH-Responsive Membranes

4.1 INTRODUCTION

The polymeric membranes are the most commonly prepared, used, and researched membranes on account of their properties, such as ease of preparation, wide availability of material, cost effectiveness, and vast applicability. Polymeric membranes are successfully used in applications, such as desalination, textiles, food, pharma, biotech, sugar, and the tannery industry. The polymeric membranes selectively separate solutes present in the feed and, depending upon the desired product, either permeate or retentate what is collected and processed. There are many transport mechanisms governing the transport of solutes across polymeric membranes, such as diffusion, convection, and sieving. Out of these, diffusion is prominent, especially solution and Knudsen diffusion.

Recently, stimuli-responsive membranes have been developed to exert more control over the permeability and separation of the membranes. The stimuli-responsive membranes show a change in performance or characteristics in response to an external stimulus, such as pH, concentration, temperature, pressure, light, magnetic, ultrasound, or electrical. Many of the external stimuli are not of much practicality and, therefore, only a few, such as pH, temperature, light, electrical, and magnetic, are suitable in various industrial applications of membrane processes. A very common trait that is easily available in different materials and applications is pH sensitivity. Therefore, it would be very beneficial to have pH-responsive membranes (PRMs). Altogether, it was the first external stimulus to be explored and it is used for making drug delivery systems more effective and efficient. In this chapter, details regarding PRMs are covered starting from the basics and progressing to their various types and applications. In-depth details are provided about the methods by which PRMs, in two configurations, flat sheet and hollow fiber, may be prepared, including blending, grafting, pore-filled method, and surface coating. The advantages and disadvantages of these methods are also discussed in this chapter.

4.2 SYNTHESIS TECHNIQUES

4.2.1 FLAT SHEET pH-RESPONSIVE MEMBRANES

Membranes with a flat sheet configuration are the most widely prepared membranes at the lab scale because of their simple and easy preparation procedures and techniques. Therefore, it can be said that they are the foundation stones for further

DOI: 10.1201/9781003201014-4

developments for a particular type of membrane. Their role is crucial for cementing the membrane field in this present era of advanced research and development. The main entity required for the preparation of flat sheet PRMs is a pH-responsive group or polymer. These pH-responsive groups or polymers are added to the membranes either by blending them with a membrane casting solution or grafting them over the membrane surface, inside the pores, or filled in the pores. The main criterium for these groups or polymers to impart pH responsiveness to the membrane is their surface presence, that is, they should be present on the membrane or membrane pore surface. Otherwise, their mere presence in a membrane is of no use when pH responsiveness is considered, even if they are uniformly distributed throughout the membrane. Therefore, it is important to keep them on the membrane surface for better results. In this section, we discuss the methods of imparting pH responsiveness to membranes.

4.2.1.1   Blended pH-Responsive Membranes

The blending method of membrane preparation is the most significant, as well as the simplest of the available methods. The major advantage of this method is that the additives used are firmly and uniformly distributed in the membranes. Also, there is less chance of leakage of the additives from the membrane over time. Therefore, these traits of the blending method make it one of the most widely used and trusted methods for the addition of different additives in membranes. In one study, the simple method of blending carboxyl groups containing the polymer poly(acrylic acid) was used by Wei et al. [1] for the preparation of flat sheet PRMs. In this study, a poly(acrylic acid) gel, prepared via solution polymerization of acrylic acids in dimethylacetamide, was blended with polyethersulfone membranes to impart pH responsiveness to the prepared membranes. The prepared membranes showed the pH valve effect in between pH 3 and 8. This was confirmed by pH-based water flux studies, which showed that the water flux at pH 2.3 was 138 and 40 mL/hm^2 mmHg at pH 8.6. This is due to the fact that the pK_a value of the additive poly(acrylic acid) lies between 4.3 and 4.9. Therefore, at lower pH, the carboxyl groups present in the additive are in their unionized state, resulting in an increase in pore size and, at higher pH, dissociation and extension of the carboxyl groups takes place resulting in a decrease in pore size. The membranes were also studied for their ion-exchange capacities with Cu^{2+} ions. The membranes had Cu^{2+} ion adsorption of 99.6%, which shows that the prepared membranes could be used as ion-exchange membranes.

PRMs are pivotal for ion-exchange applications of membranes. M'Bareck et al. studied this property of PRMs by blending poly(acrylic acid) to polysulfone flat sheet membranes and examining their application for water treatment [2]. The prepared membranes showed an ion exchange capacity in the range of 0.70–1.0 meq/g. Also, the retention and rejection of lead ions (Pb^{2+}) and Ponceau S dye (a water pollutant) by the membranes was 99% and 90%, respectively, at only 0.5 bar along with 18 L/m^2 h average pure water flux. These results show that the prepared membranes are good for water treatment applications.

In a recent study, flat sheet polymeric membranes were prepared by using poly (acrylic acid-co-polyethylene glycol methyl ether methacrylate), a novel cross-linked pegylated functional copolymer [3]. The copolymer was synthesized from acrylic acid and poly(ethylene glycol methyl ether methacrylate) using precipitation polymerization. The direct blending method was used for the addition of the additive to the polysulfone membranes and the phase inversion method was used to prepare the flat sheet membranes using N-methylpyrrolidone as a solvent. The membranes blended with the additive showed increased pure water flux, hydrophilicity, and pore density as compared to the unblended membranes (without additive). The membranes showed pH responsiveness because the hydraulic permeability of the blended membranes increased from 0.355 to 0.794 L/m^2 h kPa with pH variation from 11 to 2. Also, the antifouling property of the membranes was analyzed with BSA protein rejection. The blended membranes showed a 90% flux recovery ratio after BSA separation as compared to 37% with the unblended membranes. The BSA rejection was also studied with respect to the pH responsiveness of the prepared membranes. The study was carried out at pH 2, 7, and 11. The results of the study showed that the blended membranes have equivalent BSA rejections at pH 7 and 11, with very low rejections at pH 2. This is due to the shrinkage of the additive at lower pH values, which results in pore openings in the membranes. The mechanism of pH responsiveness has already been shown in Chapter 2, Figure 2.3. The increased pore size allows the passage of BSA across the membrane; hence, there is lower rejection at lower pH values. This is also the reason for higher hydraulic permeability at lower pH values. This confirms the pH responsiveness of the prepared membranes.

Composite membranes have also been studied in a flat sheet configuration, along with their pH responsiveness. Amphiphilic block copolymers and polystyrene-block-poly(acrylic acid) have been blended with the poly(ethersulfone) composite membranes to impart pH responsiveness [4]. The composite membranes were tested for their pH responsiveness by switching the pH between pH 3 and 8 with FITC-dextran (70 kDa). The sieving of FITC-dextran at pH 3 and 8 resulted in 28% and 90% rejection, respectively, by the prepared composite membranes. The variation in the rejection was due to the pH responsiveness of the composite membranes. This is because at pH 3 ($<pK_a$ of poly(acrylic acid)), due to the shrunken state of the poly(acrylic acid) chains, the membrane pores are in an open state and, whereas at pH 8, they are in a closed state because of the presence of poly(acrylic acid) chains in a swollen state at the membrane pore surfaces. A schematic of the membrane fabrication and pH-responsive gating function is shown in Figure 4.1. This study provides subtle evidence that PRM can also be prepared by blending amphiphilic block copolymers.

These studies with copolymers further show that there is a plausible chance of producing membranes with not only pH responsiveness but also with other crucial properties, such as hydrophilicity, antibacterial properties, and antifouling properties. These properties can be imparted to a membrane along with pH responsiveness by using copolymers consisting of polymers having different (desired) properties.

4.2.1.2 Surface-Grafted pH-Responsive Membranes

Surface-grafted PRMs are prepared by the surface graft polymerization method, in which pH-responsive groups or polymers are grafted on the membrane and membrane pore surfaces for better response. There are many methods by which the

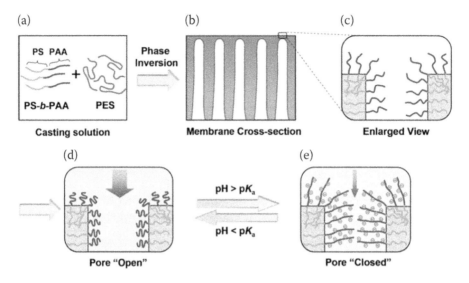

(a)   PS PAA   Phase
                                Inversion

       PS-*b*-PAA   PES

       Casting solution

(b)   Membrane Cross-section

(c)   Enlarged View

(d)   Pore "Open"

                               pH > pK$_a$

                               pH < pK$_a$

(e)   Pore "Closed"

FIGURE 4.1 Schematic illustration of the fabrication process and the pH-responsive gating function of the proposed membrane. Amphiphilic polystyrene-block-poly(acrylic acid) (PS-b-PAA) copolymers are blended with poly(ethersulfone) (PES) (a) to fabricate asymmetric PES composite membrane (b) with PAA blocks segregated on the membrane pore surface (c). The surface-segregated PAA chains on the membrane pore surfaces shrink at pH < pKa and thus induce the "open" state of membrane pores (d) and swell extensively at pH 4 pKa and thus result in the "closed" state of membrane pores (e) (reproduced with permission from Luo et al. [4] © Elsevier).

surface grafting of pH-responsive groups or polymers can be performed on the membrane surface, such as Atom Transfer Radical Polymerization (ATRP); Reversible Addition-Fragmentation chain-Transfer polymerization (RAFT); redox grafting; grafting in supercritical carbon dioxide; and the temperature-, photo-, ozone-, and plasma-induced grafting polymerization techniques, which may be "grafting-to" or "grafting-from" methods. The "grafting-to" method is the method in which the functional groups or polymers are directly grafted to the surface. In the case of "grafting from," the grafting reaction of polymerization starts from the surface that is the functional groups' or polymer chains' link to functional chains on the surface. The different methods of surface grafting are explained in the succeeding sections.

4.2.1.2.1 Atom Transfer Radical Polymerization

Atom Transfer Radical Polymerization (ATRP) is a technique for the preparation of polymers by using reversible-deactivation radical polymerization [5]. It was independently discovered by Krzysztof Matyjaszewski and Mitsuo Sawamoto in 1995. The atom transfer, as its name suggests, is important and plays a crucial role in the uniform growth of a polymer chain. There are four components of an ATRP reaction system, namely, monomer, initiator, catalyst, and solvent. In general, it implies a transition metal complex as a catalyst and a halide as an initiator. In this process, the transition metal complexes activate a dormant species for the generation of free radicals. Meanwhile, the transition metal is oxidized to a higher oxidation state. ATRP is a reversible process and attains equilibrium instantly, which is inclined toward the side with low radical concentrations. ATRP results in the formation of polymers with a narrow molecular weight range due to the tendency of each chain to grow with a monomer, because the number of initiators defines the number of chain growths [6]. Figure 4.2 shows schematically the process of surface polymerization and grafting using ATRP.

This method of polymerization is both efficient and effective. Also, it is useful because the catalysts, ligands, and initiators required for it are very simple to prepare, easily available, and low priced. ATRP is a very strong technique because it can handle a variety of functional groups present either in the monomer or the initiator, such as vinyl, allyl, hydroxyl, epoxy, and amino. PRMs are prepared by using the ATRP technique via "grafting-to," "grafting-from," and "grafting-through" methods by taking into account the grafting polymers. ATRP is used to prepare dual-responsive membranes by first activating the membrane surfaces with an alkyl halide initiator and then consecutively attaching the pH- and temperature-responsive groups. The pH-responsive groups, N,N0-dimethyl aminoethyl methacrylate [7], acrylic acid [8], and poly[(2-(diethylamino)ethyl methacrylate] [9], are

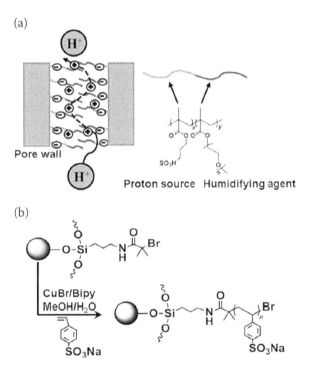

FIGURE 4.2 (a) Schematic illustrations for the pore-filling surface polymerization used to create the proton conducting channels; and (b) schematic illustration for inorganic particle surface grafting (reproduced with permission from Ran et al. [6] © Elsevier).

used along with N-isopropylacrylamide, a temperature-responsive material, for the preparation of pH- and temperature-responsive membranes by using surface-initiated ATRP.

The prepared membranes show excellent pH response, which is found to be reversible and not affected by the graft sequence. The water permeability of the membranes changes drastically between pH 6 and 8, and temperature 30–35°C.

4.2.1.2.2 Reversible Addition-Fragmentation Chain-Transfer Polymerization

Reversible addition-fragmentation chain-transfer polymerization (RAFT) is another reversible-deactivation radical polymerization technique, similar to ATRP, but with more control over the polymerization, such as a predefined and narrow range of molecular weights. It was discovered in 1998 at the Commonwealth Scientific and Industrial Research Organization (CSIRO), Australia. It can be used to prepare polymers with a variety of complex structures, such as block copolymers; dendrimers; cross-linked networks; and star-, comb-, or brush-like polymers. Thiocarbonylthio compounds, such as dithioesters, thiocarbamates, and xanthates, are chain transfer agents commonly known as RAFT agents, and required for RAFT polymerization. The RAFT agent is responsible for the control of molecular weight and polydispersity of a polymer in RAFT polymerization. RAFT polymerization is also a very tolerant technique and thus can be carried out over a wide range of temperatures and functionalities in a monomer and solvent. In general, a RAFT system comprises a monomer, radical source, RAFT agent, and solvent. A suitable RAFT agent in an appropriate amount, along with a specific temperature, is chosen so that the chain growth delivery of the free radical occurs at the required rate and the RAFT equilibrium is maintained toward the active state compared with the dormant state. This overwhelming control over the polymerization process makes it a powerful tool, suitable for the synthesis of a variety of materials with different properties. RAFT polymerization is used for the preparation of dual-responsive membranes showing responsiveness to pH as well as temperature changes [10,11]. The synthesis of the responsive copolymer by RAFT-mediated graft polymerization, along with the fabrication of pH- and thermo-responsive membranes, is shown in Figure 4.3. In these studies, membranes are grafted with poly(acrylic acid) for pH responsiveness. Also, temperature responsiveness is achieved by means of grafting poly(N-isopropyl acrylamide) onto the membranes.

These studies confirmed that by using RAFT, both pH- and thermo-responsive membranes can be prepared successfully. Also, the degree of grafting depends upon the monomer concentration and pre-treatment exposure time, such as UV and ozone treatment in [10,11], respectively. Also, it has been mentioned that the modified membranes can be further functionalized via RAFT polymerization to get multi-responsive membranes, so long as the membrane has live transfer agents present on its surface.

4.2.1.2.3 Redox Grafting

Redox grafting is a convenient, economical, and mild surface graft technique. This technique has been successfully used for the modification of ultrafiltration [12], nanofiltration [13], and reverse osmosis [14] membranes so as to improve their

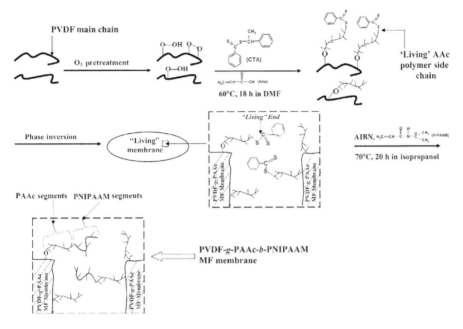

FIGURE 4.3 Schematic illustration of the synthesis of PVDF-g-PAAc copolymer by RAFT mediated graft polymerization, the preparation of the PVDF-g-PAAc membrane with living surfaces, and the preparation of the pH- and temperature-sensitive PVDF-g-PAAcb-PNIPAAM MF membrane via the surface-initiated block copolymerization (obtained from Ying et al. [11] © American Chemical Society).

antifouling properties. The redox grafting method has been used for the preparation of PRM from a novel polymeric material (membrane material), phenolphthalein poly(ethersulfone), and methylacrylic acid (functional monomer) [15]. The redox system potassium persulfate ($K_2S_2O_8$)-sodium sulfite (Na_2SO_3) was used for the initiation of the grafting process. Freshly prepared membrane samples of specific size (5 cm × 5 cm) were dipped in a solution containing quantified amounts of methyl acrylic acid and sodium sulfite. The pH of the solution was set at pH 2 by using sulfuric acid. On completion of the degassing process, the initiator potassium persulfate was added to the solution containing membrane samples at 25°C. The membrane samples were taken out of the solution after 6 h and washed thoroughly with water at 60°C for 36 h. This helps in removal of the excess and residual monomers and homopolymers. The grafting of methylacrylic acid over the membrane surface and the morphological changes in the membrane were analyzed by using Attenuated total reflection-Fourier transform infrared spectroscopy (ATR-FTIR) and field emission scanning electron microscope, respectively. Further, hydraulic permeability and diffusional experiments (with KCl and VB12) were carried out to assess the pH responsiveness of the prepared membranes. The results of this study showed that the prepared membranes presented a clear, swift, and reversible pH response. This confirms that it is possible to use the redox grafting method for the preparation of PRMs. Redox grafting, due to its various properties, is a promising, prominent, and versatile technique for the surface modification of membranes or to make them pH responsive.

4.2.1.2.4 Grafting in Supercritical Carbon Dioxide

The unique properties of supercritical fluids make them desirable for many processes. Supercritical carbon dioxide is the most useful because of its properties, such as it being economical, nontoxic, and nonflammable. Also, its critical parameters are reasonable, with a critical temperature of 31.1°C and a critical pressure of 7.38 MPa. Recently, the major use of critical carbon dioxide is as a solvent in the fields of polymer processing and chemistry. It is used to develop various polymers, fine particles, fibers, polymer fractionation, and foam materials. Supercritical carbon dioxide has the ability to dissolve smaller organic compounds and swell many others. Therefore, supercritical carbon dioxide is useful for the impregnation of different additives in polymers. The properties, such as high diffusivity, low viscosity, negligible surface tension, and tunable solvent strength, help in the uniform distribution of the monomers along with initiators in a very short period of time. Also, it is easy to remove carbon dioxide from the final product because it will be in the gaseous state at ambient conditions of pressure and temperature. Thus, it is very effective and beneficial to use supercritical carbon dioxide for the modification of membranes.

The grafting of acrylic acid onto polypropylene by using supercritical carbon dioxide was studied by Wang et al. [16] for the preparation of PRMs. The effects of temperature, pressure, and monomer concentration were studied and it was concluded that the degree of grafting can be easily tuned by regulating these parameters. The water flux of these membranes decreases drastically with a change in pH from 3 to 6. This shows that the prepared membranes change with a change in pH and hence are pH responsive. In a similar study, Gang-sheng et al. [17] studied the cross-linking and gel formation in the preparation of acrylic-acid-grafted polypropylene membranes by using supercritical carbon dioxide. The study also showed that the grafted membranes were thermally more stable compared to the nascent membranes. Again, the temperature, pressure, and monomer concentration played a vital role in the degree of grafting and gel formation. Therefore, it is very important to control these parameters for better output. Supercritical carbon dioxide has also been studied as a solvent and swelling agent for the grafting of glycidyl methacrylate over polypropylene membranes using free radical grafting [18]. The polypropylene fibers were first soaked in a solution of glycidyl methacrylate, benzoyl peroxide (initiator), and supercritical carbon dioxide is the solvent. The solvent parameters, such as temperature, pressure, and thermal treatment time, were varied to analyze the degree of grafting and morphology of the polypropylene fibers. The grafting of glycidyl methacrylate molecules onto the polypropylene fibers was confirmed by using Fourier transform infrared spectroscopy. Further, the X-ray diffraction data showed that the higher the degree of grafting, the lower was the crystallinity of the membranes. Further, supercritical carbon dioxide has been used for the free radical grafting of acrylic acid onto isotactic polypropylene while using styrene as a comonomer [19]. Similar to the other studies discussed previously, here also the effect of reaction parameters, such as temperature and pressure, on the degree of grafting were studied. The study showed that with an increase in the reaction temperature, the diffusion of monomers and radicals also increase in the supercritical carbon dioxide reaction system. Also, the increase in temperature increased the decomposition of the initiator Azobisisobutyronitrile (AIBN), resulting in increased grafting. On the other hand, the pressure of the reaction system had the opposite effect on the degree of grafting, that is, with an increase in the pressure of the reaction system there was a decrease in the degree of grafting. The comonomer styrene also plays a substantial role in the degree of grafting of acrylic acid onto isotactic polypropylene because it reacts much faster than the acrylic acid with the isotactic polypropylene radicals. Therefore, styrene reacts with the isotactic polypropylene radicals to form stabilized styrene microradicals, which later copolymerize the acrylic acid to form branches. The concentrations or amounts of initiator and comonomer used are also important for a favorable reaction and to obtain the maximum degree of grafting. For example, in the study the optimum amount of the initiator AIBN was found to be 0.75 wt%. The initiator, in quantities more than 0.75 wt%, results in a decreased degree of grafting because an excess of the initiator results in degradation of the isotactic polypropylene backbone. Further, it was confirmed that the higher the degree of grafting, the lower will be the degree of crystallinity because the grafted acrylic acid branches disrupt the regularity of the isotactic polypropylene chain. Therefore, it can be said that use of supercritical carbon dioxide in the preparation of PRMs is both effective and economical.

4.2.1.2.5 Temperature-Induced Grafting

The temperature-induced grafting method is a simple and easy-to-use method for the grafting of pH-responsive groups or polymers over a membrane surface. The basic requirement of this method is a chemical initiator or a cleavage agent, such as Azobisisobutyronitrile (AIBN). Generally, the polymerization is performed in a heterogeneous polymer-monomer reaction system in a solvent, such as water, toluene, ethanol, and others. Later, this prepared pH-responsive polymer is used to synthesize PRM. Both the "graft to" and "graft from" routes can be used. Sinha et al. [3] used this method for the preparation of pH-responsive flat sheet polysulfone ultrafiltration membranes. In this study, the copolymer poly(acrylic acid)-co-poly(ethylene glycol methyl ether methacrylate) (poly(AA-co-PEGMA)) was prepared by precipitation polymerization in toluene using AIBN as the chemical initiator. The reaction components were taken in a round-bottom three-neck flask having a condenser and an inlet and outlet for nitrogen gas. The reaction was carried out in an inert atmosphere in a boiling oil bath for ~2 h. On completion of the reaction, the reaction mixture was brought to room temperature by cooling and then vacuum filtered. Lastly, the final product was heat dried at 50°C for 4 days. The presence of pH-responsive groups in the copolymer was checked by using Fourier transform infrared spectroscopy (FTIR). Later, this pH-responsive copolymer was blended in the polysulfone membranes to impart pH responsiveness and tested for pH response with pH-based bovine serum albumin rejections.

In a similar study, polypropylene (PP) membranes were made pH responsive, but by using the "grafting-from" technique [20]. The dried PP fibers (~0.06 g) were dipped in toluene along with 0.5% benzoyl peroxide (BPO) and 1 wt% poly(vinyl

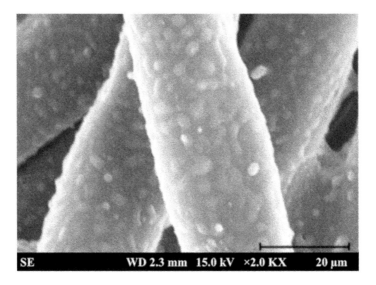

FIGURE 4.4 The scanning electron microscopy (SEM) of AA-graft-PP (reproduced with permission from Duann et al. [20] © Elsevier).

acetate). The BPO-coated PP fibers were then dried at room temperature under reduced pressure. Later, the dried BPO-coated PP fibers were immersed in a solution of water and acrylic acid (AA) with a ratio of 3:1 and 25% (w/w). A round-bottom flask with three necks was used for this reaction with a condenser. This reaction mixture under inert atmosphere was heated to 80°C for 4 h. The final product was washed with warm water so as to remove any remaining homogeneous polymer and then vacuum dried at 40°C for 24 h. The modified PP fibers showed a graft percentage of AA as ~3,200. Again, FTIR spectroscopy was used for the confirmation of the successful grafting of AA on PP fibers. The final AA-g-PP membranes are shown in Figure 4.4.

In a similar fashion, 3-(sulfopropyldimethylamino)ethyl-2-methyl acrylate (SAMA) was also grafted on PP fibers in this study. Figure 4.5 shows the reaction schematics for the synthesis of SAMA.

The study carried out by Wang et al. [21] estimated the effect of temperature, monomer concentration, and grafting time on the degree of grafting of AA on PP fiber surface. Also, they employed the efficient acid-base titration method for the estimation of the degree of grafting instead of the conventional gravimetric method. Furthermore, the chain length of the grafted AA on each grafting site of PP fiber was estimated by using high-performance liquid chromatography. In addition to this, the mechanism of grafting was also proposed.

4.2.1.2.6 Photo-Induced Grafting
Photo-induced grafting is a method of grafting pH-responsive groups or polymers on membrane surfaces in which light proliferates the polymerization reaction. It is a simple-to-use, widely accepted, and versatile method for surface grafting, not only for pH responsive, but also for thermoresponsive and electroresponsive groups or

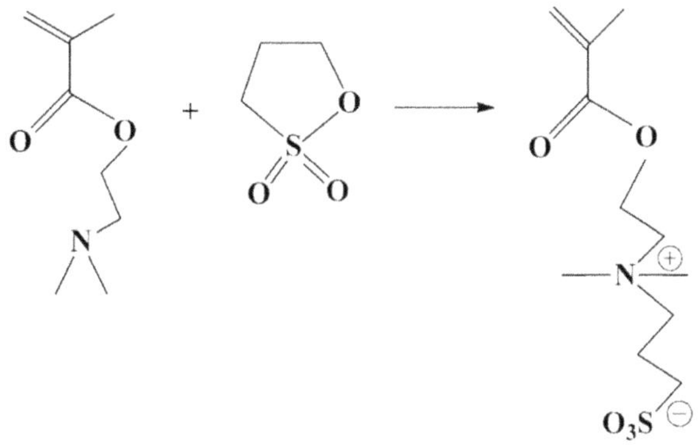

FIGURE 4.5 The synthetic process of 3-(sulfopropyldimethylamino)ethyl-2-methyl acrylate (SAMA) (reproduced with permission from Duann et al. [20] © Elsevier).

polymers, or other functional groups or polymers. Generally, UV light is used as the energy source for the reactions. Various studies have been carried out by using UV light as the mode of surface grafting pH-responsive groups or polymers on the membrane surfaces. In addition to UV light, gamma radiation has also been used by some researchers for this purpose. Further, photo-induced surface grafting can be achieved with and without photoinitiators. These are explained below.

Photo-Induced Surface Grafting with Photoinitiators   In this method, a photoinitiator, such as benzophenone, is placed on the membrane surface and then the membranes are exposed to light in the presence of the pH-responsive group or polymer in solution form. Various researchers have explored this method for the preparation of PRMs for various applications by varying the irradiation time, monomer concentration, and photoinitiator amount, among others. This method of PRM preparation was used by Ulbricht in his study [22]. In this study, commercial nylon and polypropylene microfiltration membranes were made pH responsive. This was attained by photografting poly(acrylic acid) over the membrane surface. The membranes were first coated with the photoinitiator, benzophenone, and later irradiated with UV light in the presence of poly(acrylic acid) solution. The study tested the degree of modification of the commercial membranes and revealed that the degree of modification mainly depends on the concentration of poly(acrylic acid) solution and the irradiation time. Various other studies [23] also confirmed this fact that the degree of modification or pH responsiveness of a membrane mainly depends on the concentration of the pH-responsive group or polymer concentration and the irradiation time.

Photo-Induced Surface Grafting without Photoinitiators   Photo-induced surface grafting without photoinitiators is generally carried out with the help of cross linkers, such as N,N0-methylenebisacrylamide, which help in the adherence of the pH-responsive groups or polymers to the membrane surface via crosslinking. This is an environmentally friendly method as there is no need of any photoinitiator or requirement of pure reactants as in the case of photo-induced surface grafting with photoinitiators. This method was stated to be novel for the preparation of nanofiltration membranes by Bequet et al. [24]. Polysulfone ultrafiltration membranes and pH-responsive polymer poly(acrylic acid) were used to test the method.

Cross-linker N,N0-methylenebisacrylamide was used for the cross-linking of poly (acrylic acid) over the polysulfone membrane surface under appropriate UV irradiation. This study proved that UV is an environmentally friendly and versatile technology for the preparation of PRM. Further, Hua et al. [25] studied the preparation of PRM. In this study, Hua et al. used atom transfer radical polymerization (ATRP) for the grafting of poly(methyl methacrylate) (PMMA) over the poly(vinylidene fluoride) (PVDF) to prepare a copolymer PVDF-g-PMMA under UV irradiation. This copolymer is used as an additive for the preparation of pH-responsive microfiltration membranes. The main achievement of this study was that the polymerization occurs at room temperature and with a good conversion rate of 24% conversion achieved after just 120 min UV exposure. The molecular weight of the copolymer was tested with gel permeation chromatography and the results showed that the molecular weight of the copolymer lies in a narrow range. This shows that the UV-induced ATRP polymerization was controlled.

4.2.1.2.7 Ozone-Induced Grafting

Ozone-induced grafting is an economical, environmentally friendly, and effective membrane modification technique. Ozone-induced grafting was mainly developed for the synthesis of hemocompatibility of polymeric biomaterials because this technique uses no additional toxic chemicals or any additives [26]. In this technique, a membrane sample is subjected to ozone for making it ready for grafting a polymer onto its surface in a chamber for a given period of time at a fixed temperature. Later, after removing the excess of adsorbed ozone, the membrane sample is dipped in a solution containing the monomer to be grafted on the membrane sample. The basic mechanism and relation between peroxide concentration and ozonization time is shown in Figure 4.6. Researchers have also tried this method for the development of PRMs.

Li et al. [27] carried out a study that explored ozone-induced grafting for the preparation of polypropylene PRMs. In this study, acrylic acid grafting was carried out by peroxide initiation, that is, ozone treatment was used to first introduce peroxide onto the polypropylene membrane surface and later acrylic acid was grafted. This study explored the effects of ozone treatment and grafting conditions on the different membrane properties, such as structure, strength, degree of grafting, and water permeability. The results of the study showed that an increase in the ozone treatment time along with an increased temperature results in higher introduction of carbonyl groups and peroxide over the membrane surface. However, it was also noted that the mechanical properties of the membrane deteriorated under a higher ozone treatment time. Therefore, it is important to optimize the ozone treatment time period. Also, with an increase in grafting time, monomer (acrylic acid) concentration, and increased grafting temperature, the degree of grafting was increased. Therefore, the degree of grafting of a membrane can be tuned by altering these grafting conditions. Altogether, the number of membrane pores was reduced, as was their size, due to the grafting of acrylic acid over the membrane. This resulted in reduced water

(a)

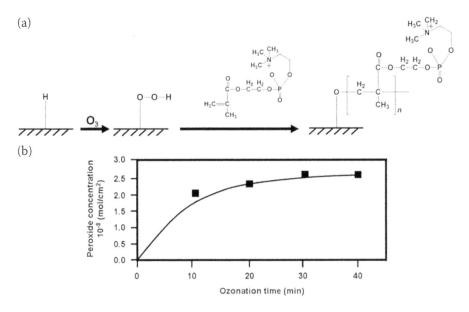

(b)

FIGURE 4.6   (a) Schematic illustration of surface graft polymerization. (b) Relationship between peroxide concentration on silicone film and ozonization time (reproduced with permission from Xu et al. [26] © Elsevier).

permeability of the grafted membrane as compared to the nascent membrane. In addition to this, the membrane water permeability also reduced with a change in pH from 3 to 6. This is due to the conformational changes of acrylic acid at these pH values. Ozone-induced grafting is thus a safe and secure grafting technique for the preparation of PRMs.

4.2.1.2.8 Plasma-Induced Grafting

Plasma polymerization can be defined as a plasma chemistry technique, involving the reaction between plasma species, plasma and surface species, or surface species, for the formation of useful polymeric materials. Polymer formation in plasma polymerization is an atomic process compared to the conventional concept of polymerization [28]. Technically, plasma polymerization follows the free radical-polymerization concept and conventionally it is similar to the free radical-polymerization of molecules having unsaturated carbon-carbon bonds [29]. In 1990, Yamaguchi et al. [30] developed a novel membrane preparation "pore filling" method using plasma polymerization. Similarly, Choi et al. produced thick and porous membranes with controlled morphologies by using plasma graft filling polymerization. The membrane was closely controlled by the plasma within the membrane pore and the gas pressure [31]. Further, Akamatsu and Yamaguchi [32] developed a new method by using plasma poly-merization for the preparation of environment-responsive membranes loaded with enzymes, as shown in Figure 4.7.

This method allows the loading of enzymes into the membrane without their denaturation and loss of activity. In this study, the "bottle in" method was used for

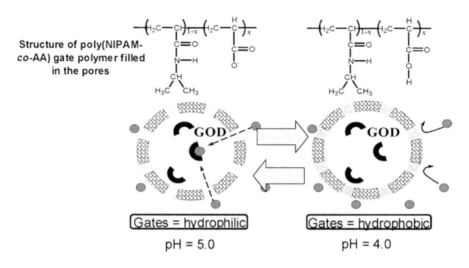

Structure of poly(NIPAM-co-AA) gate polymer filled in the pores

Gates = hydrophilic   Gates = hydrophobic
pH = 5.0   pH = 4.0

FIGURE 4.7 Schematic representation of our pH-responsive core-shell microcapsule reactor system. The microcapsule reactor is composed of a core-shell microcapsule with a porous shell membrane and a grafted pH-responsive linear polymer in the pores, with the enzymes being loaded in the core. The pH-responsive polymer grafted in the pores acts as a pH gate: the polymer is hydrated at pH 5.0 and dehydrated at pH 4.0. Thus, this microcapsule reactor has a high rate of reaction at high pH because of the high diffusivity of the solute across the hydrophilic gate polymer. Conversely, the reactor has a low rate of reaction at pH 4.0 because of the low diffusivity of the solute across the hydrophobic gate polymer, i.e., this system is a microcapsule reactor that can sense the pH of its medium and change its rate of reaction using the pH-sensitive gate polymer controlling the diffusion of the solute (obtained from Akamatsu and Yamaguchi [32] © American Chemical Society).

the loading of microcapsules with glucose oxidase and the membrane pores were filled by using plasma polymerization with N-isopropylacrylamide and acrylic acid to accomplish pH responsiveness. This resulted in the formation of pH-responsive core shell microcapsule reactors. Plasma polymerization was used by Lee and Shim [33], who synthesized acrylic acid-grafted pH-responsive poly(vinylidene fluoride) (PVDF) membranes. In this study, acrylic acid was grafted on commercial PVDF membranes using plasma polymerization. X-ray photoelectron spectroscopy (XPS) was used to confirm the grafting of acrylic acid onto the membranes. The results of this study revealed that the degree of grafting depends on the plasma exposure time. Also, the pH responsiveness analyzed with riboflavin flux showed that the permeability of riboflavin decreased drastically in between pH 4 and 5. The reason stated for this was the pK_a value of acrylic acid, that is, 4.8. Therefore, in between pH 4 and 5, carboxyl groups, present in acrylic acid, dissociated into their ionic forms and, due to ionic repulsions, exhibited extended confirmations. This change in confirmation reduced the effective membrane pore size, which resulted in decreased riboflavin permeability. Similarly, Lee and Shim [34] further prepared pH and temperature dual-responsive membranes by using the same procedure. Therefore, plasma polymerization represents a simple and effective technique for the preparation of PRM.

4.2.1.3   Pore-Filled pH-Responsive Membranes

Pore-filled PRMs are membranes with polyelectrolyte-filled pores; that is, in this type of membrane, the pH-responsive groups or polymers are present on the membrane pore surface. Therefore, the membranes show an excellent pH valve effect and proficiency in the rejection of small inorganic ions. These membrane types are important because it is the pores that control the permeability of a membrane; therefore, the presence of pH-responsive groups in the membrane pores makes the pH-responsive effect prominent. The pH-responsive effect is not very noticeable when the pH-responsive groups are present on the membrane surface or totally absent from the surface. Thus, membrane pores are the perfect place to position the pH-responsive groups. Various studies have been carried out to study pore-filled PRM. The results of these studies have confirmed that the idea is perfect for making maximum use of the pH-responsive groups. This also helps in getting membranes with a swift response to external stimuli, which is the need of various stimuli-responsive membrane applications, such as drug delivery and sensors. Mika et al. [35] prepared novel polyethylene and polypropylene microfiltration membranes by a pore-filling polyelectrolyte. 4-Vinylpyridine was successfully grafted onto the fibrils of the membranes by using the UV-induced grafting method. The membranes showed great pH-responsive behavior, as with a simple change in pH, the prepared membranes convert reversibly to reverse osmosis from microfiltration. Also, the reverse osmosis behavior of these membranes can be permanently fixed just by a simple chemical treatment.

   The pH response of the prepared membranes is totally different from membranes prepared by blending or grafting of polyelectrolytes and neutral polymers in which the response is controlled by the swelling behavior of the total membrane rather than just the polyelectrolyte phase. Reverse osmosis membranes with better performance can be prepared by cross linking methyl acrylate in a pore-filling-type membrane [36]. In this study, the plasma-induced grafting procedure was used for the grafting of methyl acrylate over the high-density polyethylene (HDPE) membrane surface. The HDPE samples were taken in a glass tube and irradiated with radio-frequency plasma (power¼ 10 W and pressure¼ 10 Pa) for 1 min in an inert (argon) atmosphere. Then, the monomer grafting was performed by placing the irradiated HDPE samples in contact with methyl acrylate aqueous solution at 10 Pa under inert (argon) atmosphere. Later, the graft polymerization was directed at 30°C in a shaking bath. Prior to the drying of the grafted membranes, they were washed with toluene overnight so as to remove the homopolymer and unreacted monomers from the membrane surface. This study confirmed that pore-filled membranes are suitable and durable for the reverse osmosis process, because the prepared membranes were stable even at high pressures up to 12 MPa. This was confirmed by the results of the reverse osmosis separation performance of the membranes with chloroform and n-hexane mixtures up to 12 MPa. Also, the membrane separation performance could be further enhanced by optimizing the crucial factors involved in membrane preparation, such as chemical structure of the cross-linkers, and the monomer and cross-linker composition.

4.2.1.4   Surface-Coated pH-Responsive Membranes

Surface coating is another technique, unlike blending and grafting, for membrane preparation. PRMs can also be prepared with this method when the coating material contains pH-responsive groups or polymers. Surface coating is instrumental in membrane preparation because it is easy to use, control, and has a high rate of success in implanting responsive groups. Therefore, it is a good technique to impart pH responsiveness to prepared membranes. Researchers around the globe have carried out research based on these surface-coated PRM. For instance, smart drug delivery membranes have been synthesized by using this method of membrane preparation [37,38]. The poly(2-hydroxyethyl methacrylate) membranes were coated with acrylic acid and 2-(diethylamino)ethyl methacrylate. The membranes carry various model drugs having different molecular weights and charges. This study showed that drug delivery, to a great extent, depends upon the interaction between the drug and the functional groups present in the membrane coating. Therefore, the results of this study will help in the design of smart release of coatings for the surface-coated membranes, especially for drugs carrying different molecular weights and charges. PRMs have also been prepared by coating layers of poly(2-vinylpyridine) [39,40], and also poly(allylamine hydrochloride) and poly(sodium 4-styrenesulfonate) [41,42]. The surface coating imparts reversible pH responsiveness to the membranes. The success rate of this method to impart pH responsiveness to the membranes makes it a great choice for membrane enthusiasts. Apart from grafting, this method also gets a good response as compared to blending because of the low efficiency of the latter.

4.3   HOLLOW FIBER MEMBRANES

Hollow fiber membranes are advantageous over flat sheet membranes in two ways. First, they are less fouling and second, they provide larger effective surface-area-to-volume ratios. At industrial scale or commercial level, hollow fiber membranes are favored as compared to flat sheet membranes because the low fouling gives them longer life and the area to volume ratio provides high packing densities. Also, it is effective and easy to use them in cross flow mode as compared to flat sheet membranes. Due to these advantages, hollow fiber membranes are playing a prominent role in the development of the membrane field in various industrial sectors. Therefore, in this section, various types of pH-responsive hollow fiber membranes are discussed based on their method of preparation.

4.3.1   BLENDED pH-RESPONSIVE HOLLOW FIBER MEMBRANES

The pH-responsive groups or polymers may be used directly for the preparation of pH-responsive hollow fiber membranes, similar to blended pH-responsive flat sheet membranes. In this case, importance should also be given to the strength and membrane-forming properties of the membrane material. Also, proper care should be taken so that the additives do not leak from the membrane.

In a recent study, Qian et al. [43] studied pH-responsive hollow fiber polyethersulfone (PES) membranes, their preparation, and their response to changes in pH, by blending a copolymer of acrylonitrile and acrylic acid (PNAA). The copolymer, PNAA, was synthesized via free radical solution polymerization by using N-methyl pyrrolidone (NMP) as solvent. Further, based on the phase inversion method, the dry-wet spinning technique was used to prepare hollow fiber membranes from a solution of PES and PNAA; again, NMP was used as the solvent. The prepared membranes showed prominent pH sensitivity and pH reversibility along with the pH valve effect between pH 4.5 and 11. Similarly, Zou et al. [38] synthesized poly (methyl methacrylate-acrylic acid-vinyl pyrrolidone) by using the free radical solution polymerization method and using dimethylacetamide (DMAC) as a solvent. The pH-responsive PES hollow fiber membranes were prepared by directly blending the functional terpolymer into the membrane casting solution using the phase inversion-based dry-wet spinning technique. The prepared membranes showed great pH sensitivity and pH valve effect between pH 7 and 10. Also, the membranes showed 10 times more water flux in acidic conditions compared to the basic environment. It was also stated that the membrane water flux variation further increased with an increase in the amount of terpolymer blended with the membranes.

The previously discussed studies related to the blended pH-responsive hollow fiber membranes prove that prominent pH-responsive hollow membranes can be prepared by direct blending of pH-responsive groups or polymers into the membranes. Further, it can be stated that these membranes carry a great potential for applications related to pH responsiveness at industrial or commercial scale. Therefore, it is a simple and economical method to consider for the preparation of pH-responsive hollow fiber membranes.

4.3.2 SURFACE-GRAFTED pH-RESPONSIVE HOLLOW FIBER MEMBRANES

Researchers have studied surface-grafted pH-responsive hollow fiber membranes as a probable source of membranes with better pH responsiveness and durability. Surface grafting is a better method to impart pH- esponsiveness to the membranes as compared to other methods because, with other methods, the problem of leakage of the pH-responsive group is there, but in surface grafting the pH-responsive group is attached to the membrane by means of cross-linking. Therefore, the pH-responsive group will remain in the membrane for longer. This makes it the method of choice for the modification of membranes.

The surface grafting method is initiated by first introducing active sites onto the membrane surface induced by light, ozone, plasma, free radicals, or temperature and then initiating monomer polymerization. For example, polypropylene (PP) hollow fiber membranes have been grafted with acrylic acid (AA) [40]. First of all, the PP membranes were washed with acetone and dried in a vacuum at room temperature. A mixture of acrylic acid and initiator benzoyl peroxide was prepared into toluene. This mixture was purged with oxygen and poured into a flask containing PP hollow fiber membranes. Then the flask was kept in an oil bath for 4 h at 60–75°C in a shaker at a speed of 150 rpm. Later, the PP hollow fiber membranes were taken out of the mixture and washed with 5% NaOH water/ethanol solution and Soxhlet extractor was used for their extraction by using methanol as solvent for 24 h. Finally, to completely remove any monomer or polymer residue, the membranes were washed with water. This is how effective and efficient AA-grafted PP hollow fiber membranes are prepared.

4.3.3 PORE-FILLED pH-RESPONSIVE HOLLOW FIBER MEMBRANES

Pore-filled pH-responsive hollow fiber membranes are sturdy, durable, efficient, and long-lasting membranes. The filling of pH-responsive groups or polymers into the membrane is done by using the grafting method because this method is effective and efficient. Pore-filled membranes are important because there is the problem of swelling, which results in the decrease of selectivity of the particular membrane. Generally, this effect can be seen with membranes separating organic liquid mixtures. Therefore, it is important to reduce this swelling of membranes for effective and efficient separation. In general, swelling in membranes can be reduced by the structure of the membranes. For example, in membranes made from polymeric alloys or emulsions, the swelling of the rubbery phase is controlled by the glassy phase.

Similarly, polyimide membranes are durable in the separation of organic liquid mixtures due to the polyimide structure. Also, swelling in membranes can be reduced by means of cross-linking and the presence of an inorganic substance in the form of a matrix in the membranes. Here, the pore-filling membranes come into effect because the swelling of the graft is reduced by the matrix of the substrate. In addition to this, the pore-filled membranes prepared from linear-grafted polymer show better permeabilities as compared to the cross-linked membranes due to the increased diffusivity of solvent in a linear polymer [42].

In a study, pore-filled porous high-density polyethylene hollow fiber membranes were prepared from the grafting of methyl acrylate into the pores of the membranes [44]. The membranes were taken into a glass tube and treated with radiofrequency plasma under inert atmosphere (argon) for 1 min at 10 Pa. Generally, water is used as a solvent in this process, but in this study 30% aqueous methanol solution was used. This is because methanol provides better homogeneous graft structures as compared to water when used as a solvent [45]. Further, the grafting process was carried out in a shaker for a fixed time at 30°C. Lastly, the grafted membranes were washed with toluene and oven dried at 40°C.

4.3.4 SURFACE-COATED pH-RESPONSIVE HOLLOW FIBER MEMBRANES

The surface coating method is a versatile method for the preparation of membranes. It provides a great scope for the customization of the preparation and development of membranes with different properties. Similarly, it is used to prepare surface-coated pH-responsive hollow fiber membranes by coating weak electrolytes on the membrane surface in a layer-by-layer fashion. This method is easy, economical, effective, and efficient. This helps in the building of a selective layer with enhanced performance and minimal solvent-induced swelling. Therefore, because of these properties, surface-coated pH-responsive hollow fiber membranes are superior to other types of PRMs.

Surface coated polyacrylonitrile (PAN) pH-responsive hollow fiber membranes were prepared by the dynamic negative layer-by-layer technique by Wang et al.

[46]. The PAN hollow fiber membranes were first hydrolyzed by immersing them into a solution of sodium hydroxide (NaOH) at 60°C. They were washed with water after hydrolyzing them for 15 min until the pH of the rinsed water reached pH 7. Later, at room temperature the polyacrylic acid (PAA) and polyetheleneimine (PEI) solutions were applied to the membranes alternatively in the inner channels of the membranes and recycled by using a peristaltic pump. On the other hand, negative pressure was created on the shell side by using a vacuum pump. The membranes were rinsed with water and purged with nitrogen gas after each cycle of poly-electrolyte deposition. In this study, the prepared membranes were post treated for 1 h with NaCl, HCl, NaOH, and NaClO by dynamically recycling the solutions. The conditions for the recycling of the solutions were maintained at 0.3 m/s recycling velocity and 25°C temperature. Lastly, the membranes were rinsed with deionized water for ~20 min and oven dried at 50°C. The performance of the prepared membranes for the pervaporation process was evaluated by using a 95-wt% ethanol and water mixture.

REFERENCES

[1] Q. Wei, J. Li, B. Qian, B. Fang, and C. Zhao. Preparation, characterization and application of functional polyethersulfone membranes blended with poly(acrylic acid) gels. *J. Membr. Sci.* 337 (2009) 266–73.

[2] C. O. M'Bareck, Q. T. Nguyen, S. Alexandre, and I. Zimmerlin. Fabrication of ion-exchange ultrafiltration membranes for water treatment. I. Semi-interpenetrating polymer networks of polysulfone and poly(acrylic acid). *J. Membr. Sci.* 278 (2006) 10–8.

[3] M. K. Sinha, and M. K. Purkait. Preparation and characterization of novel pegylated hydrophilic pH-responsive polysulfone ultrafiltration membrane. *J. Membr. Sci.* 464 (2014) 20–32.

[4] T. Luo, S. Lin, R. Xie, X. J., Ju, Z. Liu, W. Wang, C. L. Mou, C. Zhao, Q. Chen, and L. Y. Chu. pH-responsive poly(ethersulfone) composite membranes blended with amphiphilic polystyrene-block-poly(acrylic acid) copolymers. *J. Membr. Sci.* 450 (2014) 162–73.

[5] K. Matyjaszewski, and J. Xia. Atom transfer radical polymerization. *Chem. Rev.* 101 (2001) 2921–90.

[6] J. Ran, L. Wu, Z. Zhang, and T. Xu. Atom transfer radical polymerization (ATRP): a versatile and forceful tool for functional membranes. *Prog. Polym. Sci.* 39 (2014) 124–44.

[7] Z. B. Zhang, X. L. Zhu, F. J. Xu, K. G. Neoh, and E. T. Kang. Temperature- and pH-sensitive nylon membranes prepared via consecutive surface-initiated atom transfer radical graft polymerizations. *J. Membr. Sci.* 342 (2009) 300–6.

[8] K. Pana, X. Zhanga, R. Rena, and B. Caoa. Double stimuli-responsive membranes grafted with block copolymer by ATRP method. *J. Membr. Sci.* 356 (2010) 133–7.

[9] X. Qiu, X. Ren, and S. Hu. Fabrication of dual-responsive cellulose-based membrane via simplified surface-initiated ATRP. *Carbohydr. Polym.* 92 (2013) 1887–95.

[10] H. Y. Yu, W. L. J. Zhou, J. S. Gu, L. Huang, Z. Q. Tang, and X. W. Wei. Thermo- and pH-responsive polypropylene microporous membrane prepared by the photo-induced RAFT-mediated graft co-polymerization. *J. Membr. Sci.* 343 (2009) 82–9.

[11] L. Ying, W. H. Yu, E. T. Kang, and K. G. Neoh. Functional and surface-active membranes from poly(vinylidene fluoride)-graft-poly(acrylic acid) prepared via raft-mediated graft copolymerization. *Langmuir* 20 (2004) 6032–40.

[12] T. Jimbo, A. Tanioka, and N. Minoura. Characterization of an amphoteric-charged layer grafted to the pore surface of a porous membrane. *Langmuir* 14 (1998) 7112–8.

[13] S. Belfer, R. Fainshtain, Y. Purinson, J. Gilron, M. Nyström, and M. Mänttäri. Modification of NF membrane properties by in situ redox initiated graft polymerization with hydrophilic monomers. *J. Membr. Sci.* 239 (2004) 55–64.

[14] S. Belfer, Y. Purinson, R. Fainshtain, Y. Radchenko, and O. Kedem. Surface modification of commercial composite polyamide reverse osmosis membranes. *J. Membr. Sci.* 139 (1998) 175–81.

[15] M. Wang, Q. F. An, L. G. Wu, J. X. Mo, and C. J. Gao. Preparation of pH-responsive phenolphthalein poly(ether sulfone) membrane by redox-graft pore-filling polymerization technique. *J. Membr. Sci.* 287 (2007) 257–63.

[16] Y. Wang, Z. Liu, B. Han, Z. Dong, J. Wang, D. Sun, Y. Huang, and G. Chen. pH sensitive polypropylene porous membrane prepared by grafting acrylic acid in supercritical carbon dioxide. *Polymer* 45 (2004) 855–60.

[17] T. Gang-sheng, L. Taoa, Z. Linga, H. Li-xia, and Y. Wei-kang. Supercritical carbon dioxide assisted preparation of polypropylene grafted acrylic acid with high grafted content and small gel percent. *J. Supercrit. Fluids* 48 (2009) 261–8.

[18] M. H. Kunita, A. W. Rinaldi, E. M. Girotto, E. Radovanovic, E. C. Muniz, and A. F. Rubira. Grafting of glycidyl methacrylate onto polypropylene using supercritical carbon dioxide. *Eur. Polym. J.* 41 (2005) 2176–82.

[19] Q. Dong, and Y. Liu. Free-radical grafting of acrylic acid onto isotactic polypropylene using styrene as a co-monomer in supercritical carbon dioxide. *J. Appl. Polym. Sci.* 92 (2004) 2203–10.

[20] Y. F. Duann, Y. C. Chen, J. T. Shen, and Y. H. Lin. Thermal induced graft polymerization using peroxide onto polypropylene fiber. *Polymer* 45 (2004) 6839–43.

[21] W. Wang, L. Wang, X. Chen, Q. Yang, T. Sun, and J. Zhou. Study on the graft reaction of poly(propylene) fiber with acrylic acid. *Macromol. Mat. Eng.* 291 (2006) 173–80.

[22] M. Ulbricht. Photograft-polymer-modified microporous membranes with environment sensitive permeabilities. *React. Funct. Polym.* 31 (1996) 165–77.

[23] T. Peng, and Y. L. Cheng. pH-responsive permeability of PE-g-PMAA membranes. *J. Appl. Polym. Sci.* 76 (2000) 778–86.

[24] S. Bequet, T. Abenoza, P. Aptel, J. M. Espenan, J. C. Remigy, and A. Ricard. New composite membrane for water softening. *Desalination* 131 (2000) 299–305.

[25] H. Hua, Y. Xiong, C. Fua, and N. Li. pH-sensitive membranes prepared with poly (methyl methacrylate) grafted poly(vinylidene fluoride) via ultraviolet irradiation-induced atom transfer radical polymerization. *RSC Adv.* 4 (2014) 39273–9.

[26] J. Xu, Y. Yuan, B. Shan, J. Shen, and S. Lin. Ozone-induced grafting phosphor-ylcholine polymer onto silicone film grafting 2-methacryloyloxyethyl phosphor-ylcholine onto silicone film to improve hemocompatibility. *Colloids Surf. B: Biointerface* 30 (2003) 215–23.

[27] W. Li, P. Liu, H. Zou, P. Fan, and W. Xu. pH sensitive microporous polypropylene membrane prepared through ozone induced surface grafting. *Polym. Adv. Technol.* 20 (2009) 251–7.

[28] H. Yasuda. *Plasma polymerization*, Academic Press, Orlando, Florida, 1985.

[29] F. F. Shi. Recent advances in polymer thin films prepared by plasma polymerization: synthesis, structural characterization, properties and applications. *Surf. Coat. Technol.* 82 (1996) 1–15.

[30] T. Yamaguchi, S. Nakao, and S. Kimura. Plasma-graft filling polymerization: preparation of a new type of pervaporation membrane for organic liquid mixtures. *Macromolecules* 24 (1991) 5522–7.

[31] Y. Choi, S. Moon, T. Yamaguchi, and S. Nakao. New morphological control for thick, porous membranes with a plasma graft-filling polymerization. *J. Polym. Sci. A: Polym. Chem.* 41 (2003) 1216–24.

[32] K. Akamatsu, and T. Yamaguchi. Novel preparation method for obtaining pH-responsive core-shell microcapsule reactors. *Ind. Eng. Chem. Res.* 46 (2007) 124–30.

[33] Y. M. Lee, and J. K. Shim. Plasma surface graft of acrylic acid onto a porous poly (vinylidene fluoride) membrane and its riboflavin permeation. *J. Appl. Polym. Sci.* 61 (1996) 1245–50.

[34] Y. M. Lee, and J. K. Shim. Preparation of pH/temperature responsive polymer membrane by plasma polymerization and its riboflavin permeation. *Polymer* 38 (1997) 1227–32.

[35] A. M. Mika, R. F. Childs, J. M. Dickson, B. E. McCarry, and D. R. Gagnon. A new class of polyelectrolyte-filled microfiltration membranes with environmentally controlled porosity. *J. Membr. Sci.* 108 (1995) 37–56.

[36] T. Kai, H. Goto, Y. Shimizu, T. Yamaguchi, S. Nakao, and S. Kimura. Development of crosslinked plasma-graft filling polymer membranes for the reverse osmosis of organic liquid mixtures. *J. Membr. Sci.* 265 (2005) 101–7.

[37] L. T. Ng, and K. S. Ng. Photo-cured pH-responsive polyampholyte-coated membranes for controlled release of drugs with different molecular weights and charges. *Radiat. Phys. Chem.* 77 (2008) 192–9.

[38] W. Zou, Y. Huang, J. Luo, J. Liu, and C. S. Zhao. Poly(methyl methacrylate–acrylic acid vinypyrrolidone) terpolymer modified polyethersulfone hollow fiber membrane with pH-sensitivity and protein antifouling property. *J. Membr. Sci.* 358 (2010) 76–84.

[39] M. Orlov, I. Tokarev, A. Scholl, A. Doran, and S. Minko. pH-responsive thin film membranes from poly(2-vinylpyridine): water vapor induced formation of a microporous structure. *Macromolecules* 40 (2007) 2086–91.

[40] Z. Xu, J. Wang, L. Shen, D. Men, and Y. Xu. Microporous polypropylene hollow fiber membrane. Part I. Surface modification by the graft polymerization of acrylic acid. *J. Membr. Sci.* 196 (2002) 221–9.

[41] D. Lee, A. J. Nolte, A. L. Kunz, M. F. Rubner, and R. E. Cohen. pH-induced hysteretic gating of track-etched polycarbonate membranes: swelling/deswelling behavior of polyelectrolyte multilayers in confined geometry. *J. Am. Chem. Soc.* 128 (2006) 8521–9.

[42] T. Yamaguchi, A. Tominaga, S. Nakao, and S. Kimura. Chlorinated organics removal from water by plasma-graft filling polymerized membranes. *AIChE J.* 42 (1996) 892–5.

[43] B. S. Qian, J. Li, Q. Wei, P. L. Bai, B. H. Fang, and C. S. Zhao. Preparation and characterization of pH-sensitive polyethersulfone hollow fiber membrane for flux control. *J. Membr. Sci.* 344 (2009) 297–303.

[44] T. Kai, T. Tsuru, S. Nakao, and S. Kimura. Preparation of hollow-fiber membranes by plasma-graft filling polymerization for organic-liquid separation. *J. Membr. Sci.* 170 (2000) 61–70.

[45] T. Yamaguchi, S. Nakao, and S. Kimura. Evidence and mechanisms of filling polymerization by plasma-induced graft polymerization. *J. Polym. Sci. A: Polym. Chem.* 34 (1996) 1203–8.

[46] N. Wang, G. Zhang, S. Ji, Z. Qin, and Z. Liu. The salt, pH- and oxidant-responsive pervaporation behaviors of weak polyelectrolyte multilayer membranes. *J. Membr. Sci.* 354 (2010) 14–22.

5 Biomedical Applications of pH-Responsive Membranes

5.1 INTRODUCTION

The responsiveness of pH-responsive membranes (PRMs) comes about because of the presence of pH-responsive groups in the membranes. These groups are solely responsible for the pH responsiveness of PRM. In this section, some of the common pH-responsive groups are explained, along with their use in the preparation of PRM.

Generally, polymers that are weak poly-electrolytes are good pH-responsive polymers. The two most commonly used pH-responsive polymer groups are carboxyl and pyridine. A carboxyl group is alkali swellable due to the presence of carbonyl and hydroxyl functional groups. They are denoted by the chemical formula –C(=O)H and commonly written as –COOH or –CO$_2$H. These are known as alkali swellable because, at low pH, these groups protonate and hydrophobic interactions dominate among them. This leads to shrinkage of the polymer, reducing the overall volume. On the other hand, these groups form carboxylate ions at high pH due to their dissociation. This results in an enhanced charge density of the polymer, which in return increases the overall volume of the polymer due to swelling. In the case of pH-responsive polymers containing pyridine groups, the effect of pH is the opposite; that is, at a lower pH these pyridine-containing polymers swell and at higher pH values they shrink. Therefore, these groups are known as acid swellable groups. The pyridine groups protonate at lower pH and these results in an increased charge among them. Increased repulsion among the groups results and the final outcome is swelling of the polymer. In contrast, at a high pH, this charge-based repulsion among the groups decreases due to decreased ionization of the groups, thus resulting in more interactions, and hence shrinking of the polymer. Poly-(acrylic acid) and poly(methacrylic acid) are the most widely used examples of carboxyl group-containing pH-responsive polymers. On the other hand, poly(vinyl pyridine) is the most widely used pyridine-containing pH-responsive polymer. Some of the other commonly used pH-responsive groups are dibutylamine, imidazole, and tertiary amine methacrylates. A pH-based change in the pH-responsive groups is responsible for shape or configuration changes in the polymers. The pK_a values of the polymers play a vital role in pH responsiveness. In the case of the carboxyl group containing polymers, the swelling happens when the pH is higher than their pK_a values and for pyridine-containing groups it happens when the pH is lower than the pK_a value of the polymer. Therefore, based on the knowledge of the pK_a value of a polymer, its pH responsiveness can be used appropriately. Smart membranes are prepared successfully by using different types of pH-responsive polymers based on

DOI: 10.1201/9781003201014-5

their pK_a values. The desired application for a smart membrane determines the type of polymer required. Thus, it is important to have this basic knowledge about polymers for their successful use in the preparation of smart membranes.

5.2   DRUG DELIVERY

Chen et al. [1] successfully fabricated lignin-based pH-responsive nanocapsules via an interfacial miniemulsion polymerization (Figure 5.1). Lignin was first grafted with allyl groups through etherification and further dispersed in an oil-in-water (O/W) mini-emulsion system via ultrasonication. Then allyl-functionalized lignin was re-acted with a thiol-based cross-linking agent in the interfaces of miniemulsion droplets to form nanocapsules via a thiol–ene radical reaction. The FTIR and ^{1}H NMR spectra indicated the successful grafting of allyl groups on lignin. TEM images showed that lignin nanocapsules had particle sizes ranging from 50 to 300 nm. These newly synthesized nanocapsules could be readily loaded with hydrophobic coumarin-6 during the preparation of a miniemulsion system with 0.713 mmol/g entrapment efficiency. Moreover, the release of encapsulated coumarin-6 could be controlled by varying pH in the solution due to the existence of acid-labile β-thiopropionate cross-linkages in the capsule shell. An approximately linear release profile was observed at pH 7.4, whereas the release followed a Korsmeyer–Peppas profile at pH 4. The syntheses of lignin-based nanocapsules not only provide a facile approach to utilize the waste biomaterials from biorefinery industries, but also have great potential for applications in a controllable delivery of hydrophobic molecules such as drugs, essential oils, antioxidants, etc.

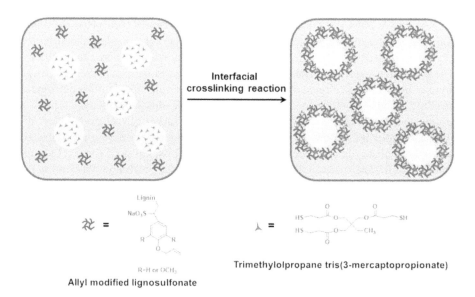

FIGURE 5.1   Preparation of lignin-based nanocapsules via interfacial miniemulsion cross-linking reaction (obtained from Chen et al. [1] © American Chemical Society).

Kim et al. [2] developed and characterized a novel intravaginal membrane platform for the pH-triggered release of nanoparticles (NPs), which was found to be essential for efficient intravaginal delivery of certain effective but acid-labile therapeutic agents for sexually transmitted infections, such as small interfering RNA (siRNA). A pH-responsive polyurethane (PU) was electrospun into a porous nanofibrous membrane which was further utilized as a semi-permeable membrane in reservoir-IVRs for pH-responsive release of NPs, as illustrated in Figure 5.2.

The diameters of the fibers, as well as the thickness and pore sizes of the membrane under dry and wet conditions (pH 4.5 and 7.0), were determined from scanning electron microscopy (SEM) micrographs. pH-dependent zeta-potential (f) of the membrane was evaluated using a SurPASS electrokinetic analyzer. VisiblexTM color-dyed polystyrene NPs (PSNs, 200 nm, ACOOH) and CCR5 siRNA-encapsulated solid lipid NPs (SLNs) were used for in vitro NP release studies in a vaginal fluid simulant (VFS) at pH 4.5 (normal physiological vaginal pH) and 7.0 (vaginal pH neutralization by semen). During 24 h of incubation in VFS, close-to-zero PSNs (2 ± 1%) and 28 ± 4% SLNs were released through the PU membrane at pH 4.5, whereas the release of PSNs and SLNs significantly increased to 60 ± 6% and 59 ± 8% at pH 7.0, respectively. The pH-responsive release of NPs hinged on the electrostatic interaction between the pH-responsive membrane and the anionic NPs, and the change in pH-responsive morphology of the membrane. In

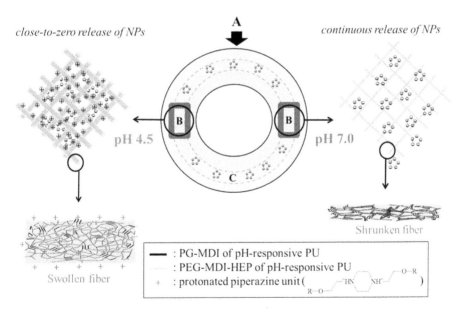

FIGURE 5.2 Diagram of the proposed use of the electrospun porous pH-responsive PU membrane as a "window" membrane in reservoir-IVR for controlled release of anionic nanoparticles release: (a) IVR, (b) window membrane, (c) reservoir. pH-responsive change in electrostatic interaction between the pH-responsive membranes and the anionic nanoparticles and morphology of the membrane contribute to the smart release of nanoparticles (reproduced with permission from Kim et al. [2] © Elsevier).

vitro biocompatibility studies of the membrane showed no significant cytotoxicity to VK2/E6E7 human epithelial cells and Sup-T1 human T-cells and no significant changes in the expression of pro-inflammatory cytokines (IL-6, IL-8, and IL-1b). Overall, the porous pH-responsive PU membrane demonstrated its potential in serving as a "window" membrane in reservoir-type intravaginal rings (IVRs) for pH-responsive intravaginal release of NPs.

Double-stimuli-responsive CNC-g-PDMAEMA reinforced PHBV electrospun nanocomposites were successfully developed by Chen et al. [3] via the electrospinning process. With the incorporation of CNC-g-PDMAEMA, the composite nanofiber membranes exhibited more smaller diameter, improved thermal stability and hydrophilicity, due to more hydrogen bonds between PHBV and CNC-g-PDMAEMA. By loading TH drugs, PHBV composite membranes exhibited better drug release behavior than those without CNC-g-PDMAEMA because of the better hydrophilicity and interaction between CNC-g-PDMAEMA nanoparticles and TH. Furthermore, drug-loaded PHBV composite membranes (6 wt% CNC-g-PDMAEMA) showed a faster drug rate at pH 5, which was beneficial for targeting the tumor cells. Thus, this study provides novel PHBV composite nanofiber membranes with double stimuli-responsive CNCs for promising systems in controlled drug release in a human physiological condition and application of molecule-targeted drugs.

Figure 5.3 demonstrates the in vitro cumulative release in PBS solutions with pH = 5, 7.4, and 8 at 37 and 45°C, respectively. Figure 5.3a,b shows the release of TH depended on the pH of the release medium, and the extent of drug release in pH = 5 PBS buffer was higher than that in pH = 8 PBS buffer. In acid condition, the charged tertiary amino group of PDMAEMA polymer-generated electrostatic repulsion between polymer chains, allowing more drug molecule to spread freely. As a result, the drug release rate was promoted. In contrast, the rate of drug release decreased with an increasing solution pH. It could be explained that the amine group got non-protonated and hydrophobic, the aggregation of polymer hindered the drug release at pH = 8 [4], as observed in Figure 5.3c.

5.3 HEMODIALYSIS

Zhu et al. [5] developed poly(lactic acid) (PLA) hemodialysis membranes with enhanced antifouling capability and hemocompatibility using poly(lactic acid)-block-poly(2-hydroxyethyl methacrylate) (PLA–PHEMA) copolymers as the blending additive. PLA–PHEMA block copolymers were synthesized via reversible addition–fragmentation (RAFT) polymerization from aminolyzed PLA. Gel permeation chromatography (GPC) and ^{1}H-nuclear magnetic resonance (^{1}H NMR) were applied to characterize the synthesized products. By blending PLA with the amphiphilic block copolymer, PLA/PLA–PHEMA membranes were prepared by the nonsolvent-induced phase separation (NIPS) method. Their chemistry and structure were characterized with X-ray photoelectron spectroscopy (XPS), scanning electron microscope (SEM), and atomic force microscopy (AFM). The results revealed that PLA/PLA–PHEMA membranes with high PLA–PHEMA contents exhibited enhanced hydrophilicity, water permeability, antifouling, and hemocompatibility, especially when the PLA–PHEMA concentration was 15 wt% and the water flux of

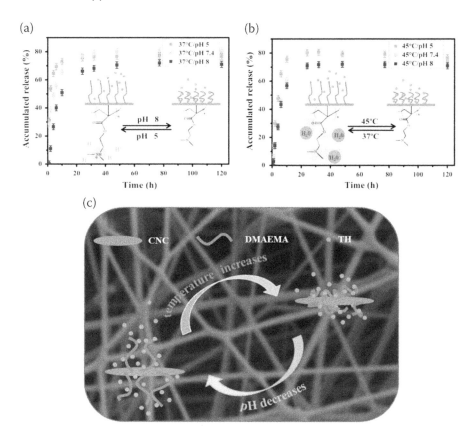

FIGURE 5.3 In vitro drug release profiles of TH drug from PHBV/CNC-*g*-PDMAEMA (6 wt%) at (a) 37°C and (b) 45°C with different pH (c) schematic illustrating possible drug release of TH from PHBV composites membranes at different physiological pH and temperature (reproduced with permission from Chen et al. [3] © Elsevier).

the modified membrane was about 236 L/m² h. Its urea and creatinine clearance was more than 0.70 mL/min, lysozyme clearance was about 0.50 mL/min, BSA clearance was as less as 0.31 mL/min. All the results suggest that PLA–PHEMA copolymers had served as effective agents for optimizing the property of a PLA-based membrane for hemodialysis applications.

During the hemodialysis process, proteins and platelets are able to absorb and deposit onto membrane surfaces leading to seriously complications. Therefore, the hemocompatibility of the hemodialysis membrane is one of the vital factors. The BSA adsorption, platelet adhesion, and plasma recalcification time (PRT) of the membranes were characterized and the results are shown in Figure 5.4. It was observed that *Adsorption*$_{BSA}$ of the membranes decreased and PRT of them increased from M0 to M20 (lower to higher additive content), indicating reduced BSA adsorption and prolonged plasma recalcification time. In addition, the pseudopodia of the adhered platelets on PLA/PLA–PHEMA membrane surfaces gradually disappeared. Meanwhile, $N_{Platelet}$ on membrane surfaces decreased from M0 to M20 (lower to higher additive content),

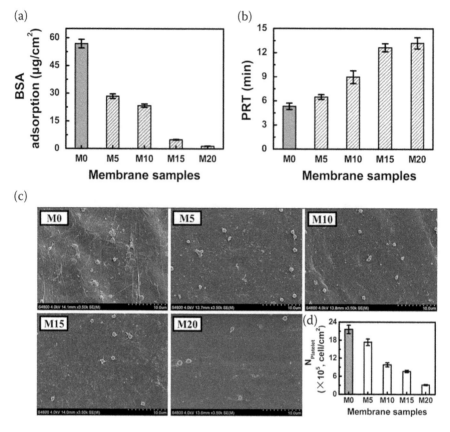

FIGURE 5.4 Hemocompatibility of the pure PLA membrane (M0) and PLA/PLA–PHEMA blend membranes (M5, M10, M15, and M20) (lower to higher additive content). (a) The amount of adsorbed BSA (Adsorption$_{BSA}$) on the membranes. Data were means ± SD ($n = 3$). (b) The plasma recalcification time (PRT) for the pure PLA and PLA/PLA–PHEMA blend membranes. Data were means ± SD ($n = 6$). (c) The typical SEM images of the adherent platelets on the membrane surfaces. (D) The number of the adherent platelets ($N_{Platelet}$) on the surfaces of membranes. Data were means ± SD ($n = 5$) (obtained from Zhu et al. [5] © American Chemical Society).

suggesting suppressed platelet adhesion and activation. All the results indicated the hemocompatibility of the PLA membrane was significantly improved by introducing PLA–PHEMA block copolymers as additives, which was mainly attributed to the improved hydrophilicity of the PLA/PLA–PHEMA membranes.

Amij [6] developed chitosan-poly(ethylene oxide) (PEO) blend membranes, using different molecular weights of PEO, for improved permeability and blood compatibility. The equilibrium hydration increased from 44.7% for chitosan to 62.5% for chitosan-PEO blend membranes when the molecular weight of PEO was 10,000 (10K) or higher. An increase in the hydration of PEO blend membranes was due to intermolecular association between PEO and chitosan chains. Scanning electron microscopy showed that chitosan-PEO membranes were highly porous, with sizes ranging from 50 to 80 nm in diameter observed in membranes made with PEO10K. Electron spectroscopy for chemical analysis suggested an increase in PEO on the membrane surface with increasing molecular weight in the blend. The permeability coefficient of urea increased from 5.47×10^{-5} cm^2/min in chitosan to 9.86×10^{-5} cm^2/min in chitosan-PEO10K membranes. The increase in permeability coefficient could be either due to an increase in the hydrophilicity or the high porosity of the membranes. Although chitosan-PEO membranes did not prevent serum complement activation, platelet adhesion and activation were significantly reduced. Chitosan-PEO blend membranes, therefore, appear to be beneficial in improving the permeability of toxic metabolites and in reducing the thrombogenicity for hemodialysis.

The use of polyethersulfone (PES)-based membranes for dialysis therapy is increasing, but the transformation and adsorption of blood proteins, destruction of red blood cells, and thrombosis responses against PES membrane can raise severe blood reactions affecting the rate of morbidity and mortality of hemodialysis (HD) patients. Irfan et al. [7] studied the performance and biocompatibility of PES membranes, which were found to improve by sulfonation and nanocomposites (NCs) additives. Acid functionalized multiwall carbon nanotubes (f-MWCNT) and polyvinylpyrrolidone (PVP) were used for NCs preparation and then they were incorporated into sulfonated-polyethersulfone (S-PES)-based membranes. The schematic chemical arrangement of sulfonated/NC-based hemodialysis membrane is shown in Figure 5.5.

The hydrophilic part of f-MWCNT contributed to the –COOH and –OH groups, whereas sulfonated polymers provided the –SO$_2$H group in the membrane composition. The formulated HD membranes were characterized by FTIR, FESEM, and contact angle. The AFM was used for the estimation of the surface roughness and surface profile studies, whereas flux rate and rejection rate were also determined. The biocompatibility results revealed that sulfonated-NC-based membranes had reduced 55% (BSA), 65% (lysozyme) adsorption, and 74.80% hemolysis process. It

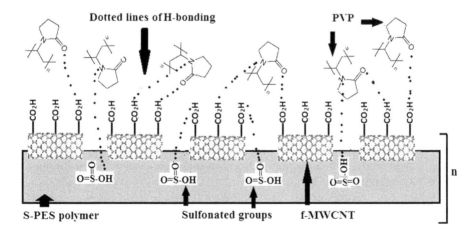

FIGURE 5.5 The schematic chemical arrangement of sulfonated/NCs-based hemodialysis membrane.

also demonstrated higher clotting time of prothrombin (PT), thrombin (TT), activated partial thrombin time (APTT), and plasma re-calcification time (PRT). The dialysis results indicated that, compared to the pristine PES membrane, the clearance ratio of lysozyme, urea, and creatinine solutes increased up to 32.4%, 59.2%, and 57.3%, respectively. Thus, the blending of S-PES and NCs in the PES membrane highly improved the biocompatibility and removal ability of uremic solutes.

5.4 ANTIBODIES AND ENZYME PRODUCTION

Membrane with pH responsiveness offers an alternative to simplify the multiple steps in purification using dual size-exclusion strategy regulated by pH switch. Melvin Ng et al. [8] synthesized a pH-responsive polysulfone (PSf) membrane to control the permeation of a monoclonal anti-ubiquitin single chain fragment variable (scFv) antibody. The change of solution pH allowed the switch of membrane's exclusion limits, retaining scFv at high pH and releasing scFv at low pH in a single unit. The small impurities could be removed from scFv at high pH while the larger impurities could be separated from scFv at low pH. The PSf membrane was incorporated with TiO_2 nanoparticles grafted with the pH-responsive polymer, poly(2-(dimethylamino)ethyl methacrylate) (PDMAEMA). FTIR confirmed that PDAMEMA was successfully grafted on TiO_2 nanoparticles. The higher ratio of DMAEMA to TiO_2 in grafting could result in the more significant protonation, improving the dispersion of TiO_2 nanoparticles. Energy dispersive X-ray analysis proved that PDMAEMA-TiO_2 nanoparticles were successfully incorporated into PSf membrane with finger-like structure as shown in scanning electron microscopic images. The molecular cut-off of membrane was improved for the pH-controlled filtration of scFv. PDMAEMATiO$_2$/PSf membrane in tangential filtration rejected scFv at pH 8 but allowed the permeation of scFv at pH 6 as proven by the sodium dodecyl sulfate polyacrylamide gel electrophoresis. By changing the buffer solution from pH 8 to pH 6, pH-responsive membrane with dual size-exclusion property could be used to purify monoclonal antibody sustainably without chromatography polishing steps.

The presence of hydrophilic TiO_2 and P-TiO_2 nanoparticles in PSf membranes improved surface hydrophilicity marginally compared to the neat PSf membrane as shown in Figure 5.6. The increment of water contact angle or hydrophobicity of PSf/P-TiO_2 5% membrane at a higher pH was caused by the deprotonation of P-TiO_2.

For the precise polishing of antibody, it is crucial to synthesis membrane with specific MWCO. PSf membranes with varied loading of P-TiO_2 4:1 nanoparticles were synthesized to determine the effects of P-TiO_2 on MWCO. As the loading was increased from 1 to 5 wt%, the MWCO of PSf/P-TiO_2 membranes was significantly refined, as shown in Figure 5.7. The refinement could be related to the reduction of particle agglomeration and the improvement of inorganic-organic compatibility in PSf/P-TiO_2 membranes.

Furthermore, Mahal et al. [9] synthesized precise morphologies of pH-responsive bioactive lead sulfide nanoparticles (PbS NPs) by using industrially and environmentally important proteins like zein and lysozyme (Lys), and a bioactive polymer diethyl aminoethyl dextran chloride (DEAE). The morphological studies of NPs' synthesis utilizing each protein have been shown in Figures 5.8–5.10.

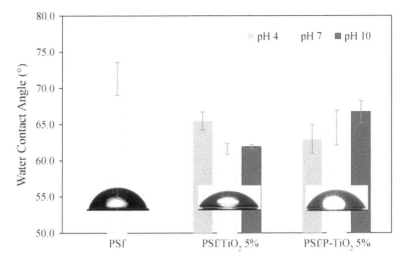

FIGURE 5.6 Water contact angle of PSf, PSf/TiO$_2$ 5% and PSf/P-TiO$_2$ 5% membranes at various pH values.

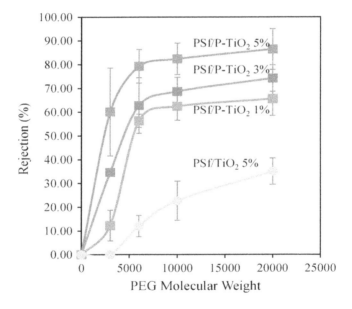

FIGURE 5.7 The effects of PDMAEMA-TiO$_2$ 4:1 nanoparticles on membrane MWCO.

Though proteins are not known as morphology control agents, zein demonstrated a fine crystal growth control of PbS NPs better than Lys as well as DEAE, and even better than conventional surfactants known for their shape control behavior. Proteins and DEAE-coated NPs thus obtained were highly pH responsive in terms of a color change from light gray (at low pH) to dark brown (at high pH). Bioapplicability of coated NPs was done by subjecting them to hemolysis. Both

(a)   (b)

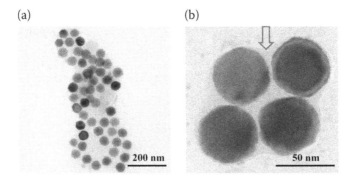

FIGURE 5.8 TEM micrographs of zein-coated PbS monodisperse spherical NPs prepared in the presence of 0.1% zein (obtained from Mahal et al. [9] © American Chemical Society).

(a)   (b)

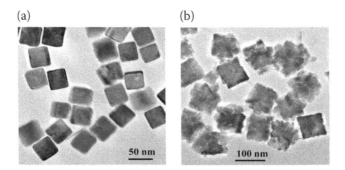

FIGURE 5.9 TEM micrographs of Lys-coated PbS cubic NPs prepared in the presence of (a) 0.4%, (b) 0.6% Lys (obtained from Mahal et al. [9] © American Chemical Society).

(a)   (b)

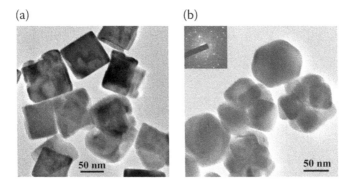

FIGURE 5.10 TEM micrographs of DEAE-coated PbS NPs prepared in the presence of 0.4% (a and b) (obtained from Mahal et al. [9] © American Chemical Society).

Lys- and DEAE-coated NPs did not induce any significant hemolysis and demonstrated their good compatibility and usability in systemic circulation. For their industrial scale uses, different extraction methods were proposed by using other industrially important biomolecules and ionic liquids. Alginic acid and xanthan gum were excellent complexing agents for an instant extraction of Lys- and DEAE-coated NPs from the aqueous phase. Ionic liquid exhibited excellent extraction ability in both organic as well as aqueous phases.

5.5 CHEMICAL SENSING AND SEPARATION

Sakaguchi et al. [10] investigated production of efficient pH-sensitive liposomes, and prepared three carboxylated poly(glycidol) derivatives with varying hydrophobicities by reacting poly(glycidol) with glutaric anhydride, 3-methylglutaric anhydride, and 1,2-cyclohexanedicarboxylic anhydride, respectively, designated as GluPG, MGluPG, and CHexPG. Previously, the group had investigated the modification with succinylated poly(glycidol) (SucPG), which provided stable egg yolk phosphatidylcholine (EYPC) liposomes with a pH-sensitive fusogenic property. Figure 5.11 shows the schematic route for synthesis of various poly(glycidol) derivatives.

Correlation between side-chain structures of these polymers and their respective abilities to sensitize stable liposomes to pH was investigated. These polymers are soluble in water at neutral pH but became water-insoluble in weakly acidic conditions. The pH at which the polymer precipitated was higher in the order SucPG < GluPG < MGluPG < CHexPG, which is consistent with the number of carbon atoms of these polymers' side chains. Although CHexPG destabilized EYPC liposomes even at neutral pH, attachment of other polymers provided pH-sensitive properties to the liposomes. The liposomes bearing polymers with higher hydrophobicity exhibited more intense responses, such as content release and membrane fusion, at mildly acidic pH and achieved more efficient cytoplasmic delivery of membrane-impermeable dye molecules. As a result, modification with appropriate hydrophobicity, MGluPG, produced highly potent pH-sensitive liposomes, which might be useful for efficient cytoplasmic delivery of bioactive molecules, such as proteins and genes.

Figure 5.12a,b depicts the time course and pH dependence of the pyranine release from the liposomes induced by the polymers. Although CHexPG induced a marked release at pH 6.5, other polymers exhibited no ability to induce a release of pyranine at any pH, indicating that SucPG, GluPG, and MGluPG have very weak or no ability to

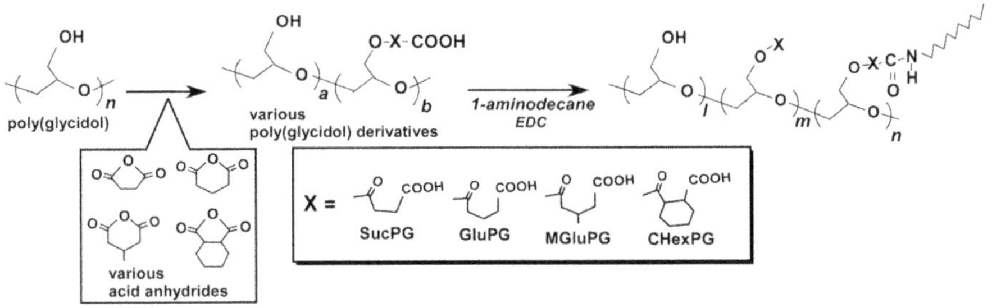

FIGURE 5.11 Synthetic route for various poly(glycidol) derivatives.[a] Scheme a EDC represents 1-ethyl-3-(3-dimethylaminopropyl) carbodiimide (obtained from Sakaguchi et al. [10] © American Chemical Society).

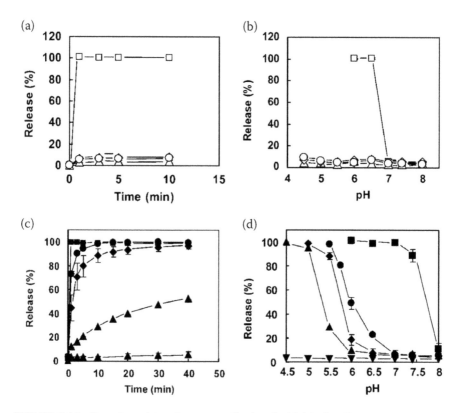

FIGURE 5.12 Pyranine release from egg yolk phosphatidylcholine liposomes induced by various poly(glycidol) derivatives. Time course at pH 6.5 (a) and pH dependence (b) of pyranine release induced by SucPG (△), GluPG (◇), MGluPG (○), and CHexPG (□). Time course at pH 6.5 (c) and pH dependence (d) of pyranine release induced by SucPG-C_{10} (▲), GluPG-C_{10} (◆), MGluPG-C_{10} (●), and CHexPG-C_{10} (■), and without polymer (▼). Measurements were performed in MES 25 mM and NaCl 125 mM solution at 25°C. Percent release of pyranine after 10 min incubation was shown (b,d). Polymer and lipid concentrations were 0.1 mg/mL and 2.0×10^{-5} M, respectively. Each point is the mean ± SD ($n = 3$) (obtained from Sakaguchi et al. [10] © American Chemical Society).

destabilize liposomal membrane. In contrast, when their counterparts with anchor moieties, namely, SucPG-C10, GluPG-C10, MGluPG-C10, and CHexPG-C10, were added to the pyranine-loaded liposomes, an intensive release of the contents was observed for all polymers (Figure 5.12c,d). Considering that most of these polymers without anchors failed to induce content release, even at acidic pH, these polymers are necessarily associated with the liposome membranes before their protonation to induce content release from the liposomes. Because their backbone with PEG-like structure can form hydrogen bonding with carboxyl groups, both the polymer backbone and side-chain carboxyl groups might not efficiently interact with liposomal membranes when the polymer chains are dissolved in solution.

Mondal et al. [11] prepared Iron nanoparticles (Fe NPs) via clove extract mediated green synthesis. Iron NPs were immobilized in flat asymmetric polymeric pH-responsive PVDF-*co*-HFP membranes prepared by blending of polyethylene glycol methyl ether (PEGME)(Mw = 5000 g/mol) and humic acid (HA). Iron NPs dispersed in pH-responsive polymeric membrane in different weight percentages (0.01, 0.1 and 1 wt%) imparted a catalytic effect in reducing nitrobenzene to aniline as well as enhanced fluoride rejection by dead-end filtration experiment. Both the process of NB reduction and fluoride rejection was dependent on pH responsiveness of the membrane. High-performance liquid chromatography (HPLC) analyzed the formation of aniline and nitrobenzene at a different ultra-filtration time.

Aniline formation of 15 ppm was found to be highest at pH 3 and lowest with 12.8 ppm at pH 7 for 0.01 wt% iron NPs impregnated membrane (optimized condition) at the end of 50 min. Aniline formation increased for pH 12 than pH 7 due to pH responsiveness of the membrane, shown in Figure 5.13. Due to deprotonation of –COOH groups present in PEGME-humic acid copolymer at a higher pH 12, the H$^+$ dominates over the membrane in comparison to pH 7; thus, aniline formation was more at pH 12 than pH 7.

Fluoride rejection was studied on optimized wt% of iron NPs (0.01 wt%) from NB reduction experiments with a different pH of fluoride solutions, as shown in Figure 5.14. Figure 5.14 explains the effect of pH on the rejection study of fluoride. The pH responsiveness of the prepared membrane plays an effective role for fluoride rejection. At pH 4, the rejection % was found to be highest for all the prepared fluoride solutions. This can be attributed to the fact that at a low pH the electrostatic interaction of the negatively charged fluoride ions increases between iron NPs and the membrane polymer. For 20, 10, and 5 ppm solutions at pH 4, the rejection was 84.4%, 78.56%, and 65.8%, respectively. With an increase in pH, the electrostatic interaction decreases, leading to a decrease in fluoride rejection. Hence, for 20, 10, and 5 ppm solutions at pH 7, the rejection decreases to 72%, 68.8%, and 55.68%, respectively.

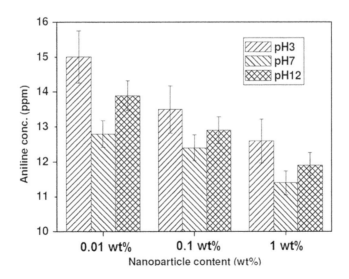

FIGURE 5.13 Overall aniline production after 50 min of NB reduction (reproduced with permission from Mondal et al. [11] © Elsevier).

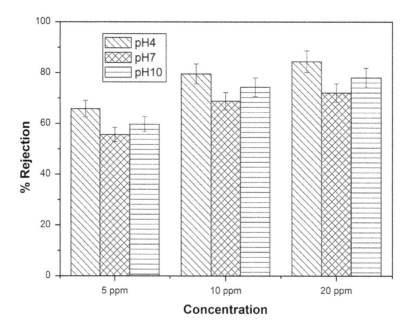

FIGURE 5.14 Rejection behavior with respect to pH and different concentration (reproduced with permission from Mondal et al. [11] © Elsevier).

But, interestingly, at pH 12 due to the de-protonation behavior of the prepared membrane, the pore size is reduced as well as H$^+$ ions prevail in the membrane polymer. This results in an increase in fluoride rejection at pH 12 but not exceeding the value for pH 4. For 20, 10, and 5 ppm solutions at pH 12, the rejection was found to be 78%, 74.25%, and 59.7%, respectively.

5.6 SELECTIVE IONIC TRANSPORT

Mondal et al. [4], through the phase inversion technique, an asymmetric flat sheet pH-responsive polysulfone (PSF) membrane was prepared and utilized for recovering H$_2$SO$_4$ in the presence of NaCl and KHCO$_3$ from wastewater. Hydrophilic and pH-responsive characteristics were incorporated within the membrane by blending polyethylene glycol methyl ether (PEGME) and humic acid (HA). The probable mechanism of cross-linking is shown in Figure 5.15.

The modification in membrane morphology with pH was characterized by field emission scanning electron microscopy (FESEM), differential scanning calorimetry (DSC), and Fourier transform infrared studies (FTIR) method. The ion exchange capacity of the prepared pH-responsive membrane increased from 0.145 to 0.25 mmol/g when compared to the pristine PSF membrane. Pure water flux (PWF) of 113.8–46.8 L/m^2 h and water uptake of 25.9–6.8% were obtained for a pH-responsive membrane when the pH varied from 4 to 12. Recovery of H$_2$SO$_4$ was optimized based on influencing parameters (pH, NaCl (M), and KHCO$_3$ (M)) through response surface methodology (RSM) and central composite design (CCD)

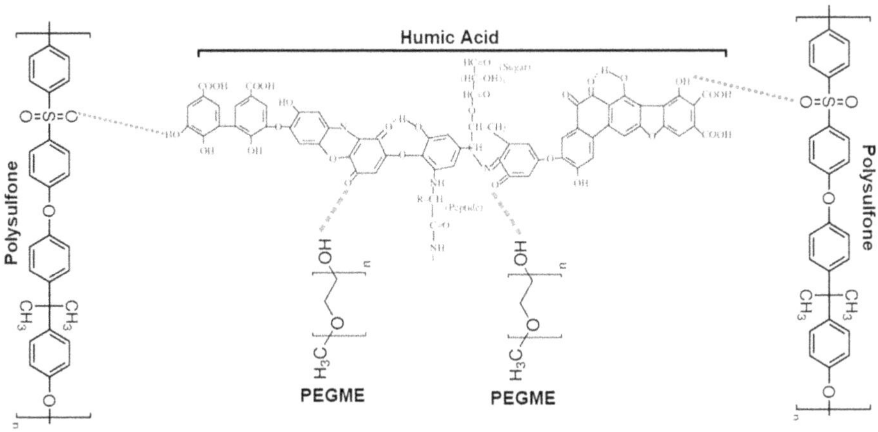

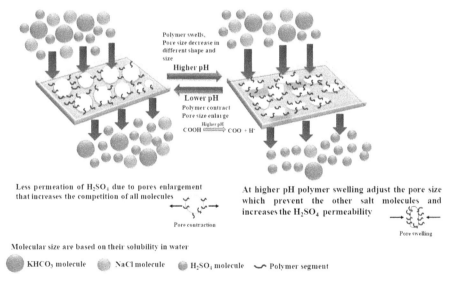

Polysulfone

Humic Acid

PEGME   PEGME

Polysulfone

FIGURE 5.15 Probable reaction mechanism of bonding between polysulfone and PEGME-co-humic acid for pH-responsive membrane preparation (reproduced with permission from Mondal et al. [4] © Elsevier).

process and was found to be a maximum of $76.57 \pm 1.5\%$ in the presence of 0.32 M NaCl and 0.5 M KHCO$_3$ at pH ~8.4, through the pH-responsive PSF membrane by diffusion dialysis process.

Figure 5.16 clearly shows that due to pore size enlargement at a low pH (pore size 68.7 nm), permeation of H$_2$SO$_4$ molecules decreases due to high competition with KHCO$_3$ and NaCl molecules (since the large pore size facilitated permeation

Polymer swells,
Pore size decrease in
different shape and
size
Higher pH

Lower pH
Polymer contract
Pore size enlarge
Higher pH
COOH ⇌ COO⁻ + H⁺

Less permeation of H$_2$SO$_4$ due to pores enlargement that increases the competition of all molecules

Pore contraction

At higher pH polymer swelling adjust the pore size which prevent the other salt molecules and increases the H$_2$SO$_4$ permeability

Pore swelling

Molecular size are based on their solubility in water

● KHCO$_3$ molecule   ● NaCl molecule   ● H$_2$SO$_4$ molecule   ∿ Polymer segment

FIGURE 5.16 H$_2$SO$_4$ recovery mechanism of acid molecules with respect to NaCl and KHCO$_3$ molecules according to pH (reproduced with permission from Mondal et al. [4] © Elsevier).

for all the molecules), whereas at a high pH, average pore size decreases to 28.8 nm, facilitating the permeation of smaller molecules such as H_2SO_4. Moreover, at an optimized condition pH ~8.4, when the membrane surface behaves negatively, the competitiveness between SO_4^{2-}, Cl^-, and HCO_3^- plays an important role in the solution for permeating through the membrane. The order of competing for an anion can be explained according to the Hofmeister Series $SO_4^{2-} > HCO_3^- > Cl^-$, thus explaining the permeation of SO_4^{2-} through the membrane. Similarly, H^+ ions are easily permeated based on its size and charge. Hence, the permeation of H_2SO_4 is a combined effect of (i) the lower molecule size of the acid, (ii) narrow pore size at a high pH for selective permeation of acid molecules, and (iii) the greater affination of SO_4^{2-} than other competing anions in the solution.

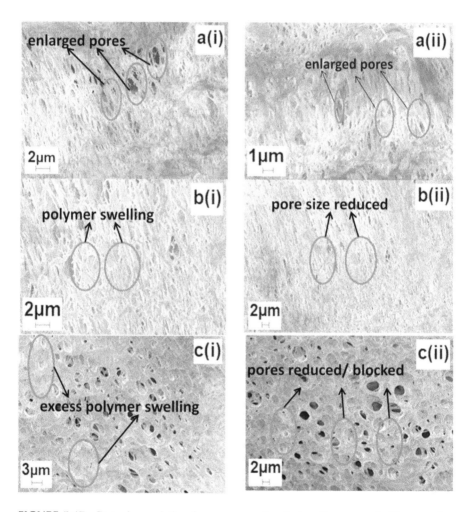

FIGURE 5.17 Pore size variation from cross-sectional view for prepared pH-responsive membranes at (a) pH 4, (b) pH 7, and (c) pH 12 (reproduced with permission from Mondal et al. [12] © Elsevier).

Similarly, Mondal et al. [12] further synthesized flat sheet, asymmetric pH-responsive poly-(vinylidene-fluoride-co-hexafluoro-propylene) (PVDF-co-HFP) membrane through the phase inversion technique and utilized for permeation study of glucose in presence of various salts.

The pH-responsive pore size changes shown in Figure 5.17 are obtained from FESEM characterization. Pure water flux (PWF) varied from 14 to 58 L/m^2 h with an increase in pH, indicating swelling and pore blocking activity of the pH-responsive membrane. MIP analysis confirmed the variation of pore size and pore distribution of the membrane at a different pH. Optimization of the variables (pH, NaCl, and KHCO$_3$) for maximizing glucose removal was performed using design expert software 9.0 TRIAL, through ANOVA (analysis of variance), using the combination of response surface methodology (RSM) and central composite design (CCD). Glucose removal by the pH-responsive membrane at a concentration of 8.3 mmol/L (higher-than-normal glucose concentration after meal), through the dialysis technique obtained a maximum of 60.67% ± 3.1% at pH 12 in the presence of a 0.1 M NaCl and KHCO$_3$ solution. At a normal human body fluid pH of 7.4, the removal was around 54.3% ± 3.5%.

REFERENCES

[1] N. Chen, L. A. Dempere, Z. Tong. Synthesis of pH-responsive lignin-based nanocapsules for controlled release of hydrophobic molecules. *ACS Sustain. Chem. Eng.* 4 (2016) 5204–11.

[2] S. Kim, Y. L. Traore, E. A. Ho, M. Shafiq, S. H. Kim, S. Liu. Design and development of pH-responsive polyurethane membranes for intravaginal release of nanomedicines. *Acta Biol.* 82 (2018) 12–23.

[3] Y. Chen, S. Y. H. Abdalkarim, H. Y. Yu Y. Li, J. Xu, J. Marek, J. Yao, K. C. Tam. Double stimuli-responsive cellulose nanocrystals reinforced electrospun PHBV composites membrane for intelligent drug release. *Int. J. Biol. Macromol.* 155 (2020) 330–9.

[4] P. Mondal, N. S. Samanta, A. Kumar, M. K. Purkait. Recovery of H$_2$SO$_4$ from wastewater in the presence of NaCl and KHCO$_3$ through pH-responsive polysulfone membrane: optimization approach. *Polym. Test.* 86 (2020) 106463.

[5] L. Zhu, F. Liu, X. Yu, L. Xue. Poly(lactic acid) hemodialysis membranes with poly (lactic acid)-block-poly(2-hydroxyethyl methacrylate) copolymer as additive: preparation, characterization, and performance. *ACS Appl. Mater. Interfaces* 7 (2015) 17748–55.

[6] M. M. Amij. Permeability and blood compatibility properties of chitosan-poly (ethylene oxide) blend membranes for haemodialysis. *Biomaterials* 16 (1995) 593–9.

[7] M. Irfan, A. Idris, R. Nasiri, J. H. Almaki. Fabrication and evaluation of polymeric membranes for blood dialysis treatments using functionalized MWCNT based nanocomposite and sulphonated-PES. *RSC Adv.* 6 (2016) 101513–25.

[8] H. K. Melvin Ng, C. P. Leo, T. S. Limb, S. C. Lowa, B. S. Ooia. Polishing monoclonal antibody using pH-responsive TiO$_2$/polysulfone membrane in dual size-exclusion strategy.*Sep. Purif. Technol.* 213 (2019) 359–67.

[9] A. Mahal, L. Tandon, P. Khullar, G. K. Ahluwalia, M. S. Bakshi. pH responsive bioactive lead sulfide nanomaterials: protein induced morphology control, bioapplicability, and bioextraction of nanomaterials. *ACS Sustain. Chem. Eng.* 5 (2017) 119–32.

[10] N. Sakaguchi, C. Kojima, A. Harada, K. Kono. Preparation of pH-sensitive poly (glycidol) derivatives with varying hydrophobicities: their ability to sensitize stable liposomes to pH. *Bioconjugate Chem.* 19 (2008) 1040–8.

[11] P. Mondal. M. K. Purkait. Green synthesized iron nanoparticles supported on pH-responsive polymeric membrane for nitrobenzene reduction and fluoride rejection study: optimization approach. *J. Cleaner Prod.* 170 (2018) 1111–23.

[12] P. Mondal, N.s. Samanta, V. Meghnani, M. K. Purkait. Selective glucose permeability in presence of various salts through tunable pore size of pH-responsive PVDF-co-HFP membrane. *Sep. Purif. Technol.* 221 (2019) 249–60.

6 Recent Trend and Future Prospects

6.1 RECENT DEVELOPMENTS IN pH-RESPONSIVE MEMBRANES

Preparation of membranes from pH-responsive polymers, copolymers, and polymer-additive mixtures is an important approach in the design of pH-responsive membranes (PRMs). This approach enables fabrication of membranes with the desired mechanical properties, pore structure (porosity, pore size, and pore-size distribution), barrier structure (symmetric versus asymmetric), and layer thickness (es). Membrane surface modification also is important in the design of PRMs, as the optimal required membrane surface characteristics rarely are achieved from membrane-forming polymers, copolymers, or polymer-additive mixtures alone. Modification imparts functionality that enhances membrane performance. By taking this approach, the useful properties of the base membrane are maintained, and re-sponsive properties are introduced to the membrane surface. When a modification is done using controlled, surface-initiated polymerization strategies such as ATRP (already discussed in Chapter 3), polymer molecular architecture can be controlled precisely, allowing fundamental studies on the role that surface architecture plays on membrane responsiveness and performance. Following are the mentioned de-velopments in the fabrication of the PRMs.

6.1.1 COPOLYMER SYSTEMS

Tamada et al. [1] prepared hydrolyzed pH-sensitive track-etched membranes from copolymer films of N-methacryloyl-l-alanine methyl ester and diethyleneglycol-bis-allylcarbonate in two steps. In the first "rigorous" etch step, copolymer films were ion-irradiated in 6 M NaOH solution at 60°C to obtain virtually cylindrical pores of 4 μm diameter. During the second "soft" etch step, the resulting filter membranes were hydrolyzed with 1 M NaOH solution to create surface layers of varying thicknesses, which were responsive to pH. The pH-responsive swelling ratio, $S = [(M - M_0)/M_0]$, was obtained from measurements of membrane mass in the swollen state (M) and mass in the dry state (M_0). The membranes showed steep increases in their swelling ratios by more than two orders of magnitude in moving from pH 3 to pH 5. The pore diameters of the membranes showed a pH response, with pore diameters of 3.7 μm in solutions of pH 3 and completely closed pores at pH 5.

Oak et al. [2] reported on the preparation of pH-sensitive UF membranes from poly(acrylonitrile (AN)-co-AA) and poly(AN-co-methacrylic acid (MAA)) by phase inversion in dimethylsulfoxide/water medium. Water permeation through the membranes was examined at a different pH, as shown in Figure 6.1.

When the pH of the permeate was varied from acid to alkaline conditions, the flux through the P(AN-co-AA) and P(AN-co-MAA) membranes decreased slightly

DOI: 10.1201/9781003201014-6

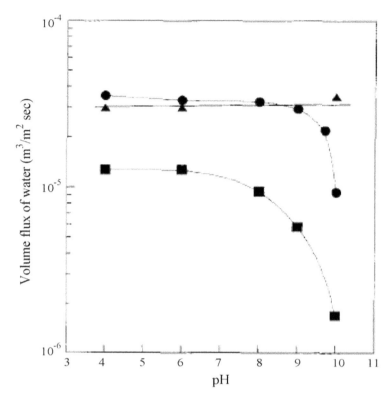

FIGURE 6.1   Water flux at various pHs under applied pressure of 760 mm H$_2$O through (▲) PAN, (●) P(AN-co-AA) copolymer, and (■) P(AN-co-MAA) copolymer membranes (Reproduced with permission from Oak et al. [2] © Elsevier).

until pH 8 and then decreased significantly in the pH 8–10 region. It was also observed that the flux through P(AN-*co*-MAA) membranes began to change gradually at pH 6, while the flux through P(AN-*co*-AA) membranes began to change in the range of pH 9–9.5. The flux through PAN only membranes remained almost constant at 3 × 10^{-5} m^3/m^2 s. Molecular weight cut-off data obtained by dextran permeation studies at pH 4, 6, and 10 showed that the pore size of the membranes decreased significantly under alkaline conditions.

6.1.2   INTERPENETRATING NETWORK SYSTEMS

Gudeman and Peppas [3,4] prepared pH-sensitive membranes with varying degrees of cross-linking from IPNs of PVA and PAA. The membranes were characterized and tested for permeation of a wide range of solutes. Changes in pH from 3 to 6 increased the membrane mass swelling ratio, accompanied by up to 86% increases in the mesh sizes. The swelling ratio also increased with a decrease in the ionic strength of the swelling medium. Membranes with loosely cross-linked structures were observed to swell and de-swell more quickly than densely cross-linked membranes. Permeation studies demonstrated that for ionizable solutes such as l-tryptophan and urea, the diffusion coefficient was smaller at pH 3 than at 6. Transport of neutral molecules depended more on the solute size.

Park et al. [5] describe organic–inorganic IPN membranes prepared using tetraorthosilicate (TEOS) as the inorganic material and chitosan as an organic compound. Chitosan ((1 → 4)-2-amino-2-deoxy-β-d-glucan) is charged positively and swells in an acidic medium and shrinks in a basic solution because of ionization of its amino groups. When incorporated in the TEOS IPN, chitosan imparts the membranes with pH sensitivity. Equilibrium swelling studies showed that the membranes swelled at pH 2.5 and shrunk at pH 7.5 regardless of TEOS–chitosan ratio. The swelling behavior was completely reversible. Also, drug permeability within the membranes changed immediately as environmental pH conditions were altered. Permeation studies showed that an increase in pH from 2.5 to 7.5 increased the rate of drug permeation regardless of TEOS–chitosan ratio, while decreases in pH resulted in low permeation rates shown in Figure 6.2. These membranes have a potential use as drug carriers and as a bioseparation platform.

6.1.3 MICRO/NANOCOMPOSITES

Turner and Cheng [6] developed composite-heterogeneous PEL gel (composite-PG) membranes consisting of PMAA gel particles dispersed within mechanically stronger, hydrophobic, elastic poly(dimethyl siloxane) (PDMS) networks and evaluated them as pH-responsive membranes. The PMAA gel particles remained responsive within the membranes allowing for external pH control of the membrane hydration and, hence, local diffusivity. The membranes with 17% and 22% PMAA gel particles showed little change in permeability in response to pH change, despite increased hydration of the gel particles in these membranes with increasing pH. This result was attributed to the low gel mole fraction of these membranes, resulting in little or no particle connectivity despite increases in particle hydration. However,

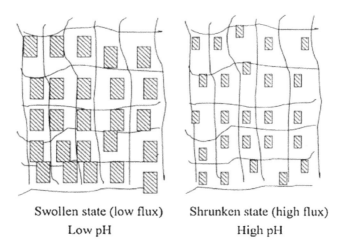

Swollen state (low flux)   Shrunken state (high flux)
Low pH   High pH

FIGURE 6.2 pH-responsive mechanism of TEOS–chitosan IPN hybrid membranes (reproduced with permission from Park et al. [5] © Elsevier).

membranes with 28% and 33% PMAA gel particle loading showed 10-fold and 40-fold increases in permeability for caffeine and vitamin B_{12}, respectively, when the pH was changed from 3 to 7.

Zhang and Wu [7] investigated the temperature- and pH-responsive permeability of various peptides, proteins, and vitamin B_{12} through composite membranes containing dispersed poly(NIPAAm-co-MAA) nanoparticles. The polymeric membranes were developed by dispersing the nanoparticles in a hydrophobic polymer (ethylcellulose). Permeability of the solutes across the membranes increased with increasing temperature but decreased with increasing pH. The temperature and pH sensitivity of the composite membranes was determined by the composition of the nanoparticles. Nanoparticles with higher percentages of NIPAAm imparted greater temperature sensitivity, while those with higher percentages of MAA imparted greater pH sensitivity to the membranes. Therefore, the membranes could be tailored for a specific application by selecting appropriate nanoparticle composition.

6.1.4 MODIFICATIONS OF pH-RESPONSIVE MEMBRANES

6.1.4.1 Grafting to Modification

Ito et al. [8] designed nanometer pore size, pH-responsive membranes by self-assembly of ionizable polypeptide brushes on gold-coated, commercial track-etched porous PC membranes. The membranes were coated with platinum and then with gold and were thereafter immersed in an aqueous solution of lyophilized poly(l-glutamic acid) (PLGA) for 24 h. Water permeation through the membranes was investigated, and it was observed that the water permeation through the unmodified membranes was independent of pH, while that through the modified membranes was dependent on pH, with a high permeation at a low pH and low permeation at near-neutral pH. At a low pH, PLGA chains are protonated and folded, forming α-helical structures that lie on the surface. At a high pH, they are deprotonated, forming extended random structures that extend into the solution. These conformational changes affect the porosity of the membrane, leading to the observed pH-dependent water permeability. The observed inflection point of the water permeation rate was at pH 4.5–5, which is the same as the isoelectric point of PLGA (4.58).

Zhang and Ito [9] synthesized self-assembled chains of PAA conjugated with cysteamines (PAA-SH) on gold-coated nanoporous membranes to produce membranes whose water permeability was pH controlled. Porous PC membranes with an average pore diameter of 200 nm were coated with gold up to a thickness of 50 nm and then exposed to an aqueous solution of PAA-SH at different pH values and ionic strength concentrations for 24 h. The modified membranes were rinsed with water until the pH of the rinse liquid became neutral. Transport through the modified membranes was investigated. The rate of water transport through the bare membrane was independent of pH, whereas the transport through the modified membranes could be regulated reversibly by variation in pH and ionic strength. High permeability was observed at a low pH, low permeability was observed at a neutral pH, and an increase in ionic strength increased the permeability at a high pH. At a high pH, permeability was strongly dependent on ionic strength but the effect was limited at a low pH. At high ionic strength, water permeability became less sensitive to changes in pH and this was attributed to the high ionic strength shielding deprotonated polymer segments from electrostatic repulsion. Filtration of solutions of ionic (oligodeoxyribonucleotide) and non-ionic PEG polymers through the modified membranes showed that pH-responsive permeability depended on the molecular weight of the solutes.

Hollman and Bhattacharyya [10] investigated the influence of covalently attached PLGA on the performance characteristics of microporous cellulosic supports. They determined the effect of polypeptide on water transport through the functionalized supports. The helix–coil transitions of the PLGA affected the permeability in a reversible fashion upon variations in solution pH. Synthesis of the functionalized cellulose membranes involved two steps: aldehyde derivatization followed by PLGA attachment. PLGA attachment was carried out by permeating aqueous PLGA solutions through the cellulosic supports at pH 9.2–9.8. Functionalization involved the reaction between the terminal amine group present on the PLGA with the aldehyde group present in the cellulosic support. The functionalized membranes displayed marked decreases in water flux at a high pH and increased water fluxes at a low pH due to the extended random-coil formation and helix formation of the PLGA chains, respectively.

6.1.4.2   Grafting from Modification

Imanishi and co-workers [11] investigated the pH-dependent changes of the pore sizes of PAA grafted, straight-pore PC membranes. Peroxide groups were generated on the surface of the PC membranes by glow discharge, and the membranes were then heated in aqueous solutions of AA to initiate the graft polymerization. Water permeability of the modified membranes increased sharply in pH regions below 4, representing an expansion of pores. However, very high grafting densities or very long grafted chains restricted the mobility of the PAA chains and the pore size became pH independent.

Ito et al. [12] developed a membrane device that controlled the rate of water permeation according to the pH and ionic strength by thermally initiated surface grafting of vinyl, carboxylic acid containing monomers from a straight-pore membrane. AA, MAA, and ethacrylic acid (EAA) were used as the functional monomers. The rate of water permeation through the modified membranes changed reversibly with changes in solution pH. The pH response of the water permeability also was controlled by changing the grafting density and the polymer layer thickness. Membranes modified with poly(carboxylic acid) of a high degree of polymerization grafted in low densities produced the most sensitive pH response. The most drastic change of water permeability occurred at pH 3.0, 4.0, and 6.8 for membranes modified with PAA, PMAA, and PEAA, respectively. These pH differences reflect different pK_a values of the grafted polymer chains: PAA, 4.8; PMAA, 6.2; PEAA, 7.2. Water permeation also was affected by the ionic strength of the aqueous solution.

Mika et al. [13] successfully synthesized composite membranes made of MF substrates and pore-filling PELs by UV-induced grafting of 4VP from PE and PP MF membranes. They showed by simple changes in pH that the barrier properties of these membranes changed reversibly from that characteristic of MF to that characteristic of reverse osmosis. At low pH values, the pyridine nitrogen atoms in P4VP are protonated to form positively charged pyridinium groups. The charged polymer has an extended chain conformation due to electrostatic repulsion, which effectively fills the pores. The flux through the membranes decreased by three to four orders of magnitude in moving from high pH to low pH. This very large change in flux occurred over a very narrow range of pH of the contacting solution, and the changes were found to be reversible.

Peng and Cheng [14] studied the pH responsiveness of PE-g-PMAA membranes. Graft yields were calculated as (Wg – Wu)/Wu, where Wu and Wg were the dry weights of the ungrafted and grafted membranes, respectively. Membranes with a wide range of graft yields were prepared by photo-grafting PMAA on porous PE membranes by UV irradiation. The pH-dependent permeability of the modified membranes was studied as a function of the graft yield. Two types of permeability response were observed, depending on the graft yield. At low graft yields, the membranes showed porous membrane-responsive behavior, that is, the collapse of the grafted polymer would leave the membrane pores open compared with an expansion of the grafted polymer. At high graft yields, the membranes became hydrogel-like, showing lower permeability in the collapsed state. Generally, it was observed that membrane permeability changed reversibly in response to solution pH alternation between pH 2 and pH 7.4.

Wang et al. [15] prepared pH-responsive membranes by grafting AA from porous PP membranes using supercritical (SC) CO_2 as a solvent. The monomer and the initiator, benzyl peroxide, were impregnated into the PP substrate with the aid of SC, CO_2 and PAA chains were then grafted from the microporous PP substrate by thermal initiation. The water permeation of the unmodified membrane was nearly independent of pH, while the water permeation of the modified membranes decreased dramatically as the pH was increased from 3 to 6 because of conformation changes in PAA.

6.2 ROLE OF pH-RESPONSIVE MEMBRANES IN SEPARATION APPLICATIONS WITH FUTURISTIC INSIGHTS

6.2.1 SIZE-EFFECT-BASED SIEVING

With the stimuli-responsive (pH responsive) self-regulation of pore size, gating membranes can be applied for graded sieving separation. Generally, only smaller molecules/particles can permeate across the membrane with closed pores, while both smaller and larger molecules/particles can permeate when the pores open (Figure 6.3a). Thus, the separation of substances with different sizes can be achieved by using single gating membranes, with the pore size regulated by designed stimuli. pH-responsive gating membranes can selectively reject dextran molecules with appropriate molecular weights from mixtures with different molecular weights of 10, 40, and 70 kDa, depending on the environmental pH (Figure 6.3b).

Similarly, temperature-responsive gating membranes enable the fast permeation of small molecules such as NaCl (hydrodynamic radius ~0.1 nm), showing large diffusional coefficients at both 25 and 40°C (Figure 6.3c); while large molecules such as VB12 (hydrodynamic radius ~2 nm) can only permeate through the

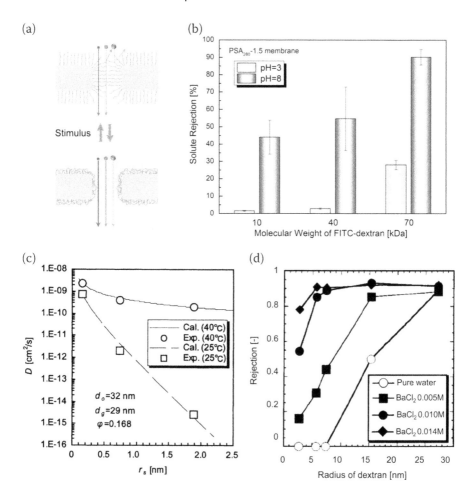

FIGURE 6.3 Smart gating membranes for size-sieving-based separation. (a) Schematic illustration of the stimuli-responsive size-sieving-based separation. (b–d) Graded size-sieving-based separations using pH-responsive (b), thermo-responsive (c), and ion-responsive (d) gating membranes (reproduced with permission from Liu et al. (2016) [28] © Royal Society of Chemistry).

membrane with opened pores at 40°C [16]. Similarly, Ba^{2+}-responsive gating membranes can sieve molecules with different sizes such as dextran molecules with radii of 2–30 nm (Figure 6.3d).

6.2.2 AFFINITY-BASED ADSORPTION/DESORPTION

With self-regulated surface properties for controlling the affinity between pore surfaces and substances, smart gating membranes offer ingenious tools for stimuli-responsive separation or purification of substances such as proteins and chiral molecules.

For example, gating membranes with gates that allow a thermo-induced switch between the hydrophilic and hydrophobic states can be used for separating

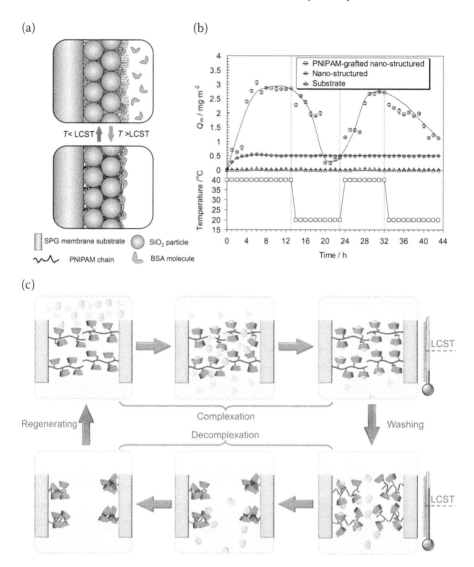

FIGURE 6.4 Smart gating membranes for affinity-based separation. (a and b) Schematic illustration (a) and experimental data (b) showing membranes with PNIPAM-based gates for thermo-induced adsorption/desorption of BSA molecules. (c) Membranes containing thermo-responsive PNIPAM chains with appended β-CD moieties as functional gates for chiral resolution (reproduced with permission from Liu et al. (2016) [28] © Royal Society of Chemistry).

hydrophobic substances such as bovine serum albumin (BSA) based on hydro-phobic adsorption (Figure 6.4a). The BSA can be adsorbed when the gates are hydrophobic, and desorbed when the gates become hydrophilic. This can be simply controlled by varying the operation temperature (Figure 6.4b). As another example, by combining PNIPAM with functional b-CD, which can act as a host molecule or chiral selector, gating membranes for chiral resolution are achieved (Figure 6.4b). At temperatures below the LCST of PNIPAM, the PNIPAM/b-CD gates are swollen and hydrophilic. During the solution permeation, one of the enantiomers can be selectively captured by the b-CD groups based on their stronger association. When increasing the temperature above the LCST, the PNIPAM/b-CD gates become shrunken and hydrophobic, leading to decomplexation of the b-CD and captured enantiomer due to the weakened association constant; thus, the enantiomer can be separated. Therefore, smart membranes with functional gates for enantioseparation allow simple membrane regeneration through changing the temperature, and show high efficiency for selective chiral resolution [17].

6.2.3 SELF-CLEANING OF MEMBRANES

Membrane fouling, which usually leads to a weakened membrane performance, such as permeability loss, is an unavoidable problem for membrane-involved processes. Generally, polymers used for porous membrane manufacture are usually hydrophobic in nature; as a result, the organic foulants in water are highly susceptible to depositing on the membrane surface due to the hydrophobic interaction between the membrane and foulants [3]. Thus, hydrophilic polymers grafted on membrane surfaces can provide stericosmotic barriers against the fouling adsorption for reduced membrane fouling; however, the grafted polymers also reduce the intrinsic permeability owing to the partial blocking of the membrane pores [3]. Smart-gating membranes with tunable surface properties create opportunities to achieve self-cleaning functionality for reducing membrane fouling while retaining the permeability. Upon adding a stimulus, the shrunken and hydrophobic gates become swollen and hydrophilic; such transitions weaken the interactions between the foulant and membrane surface for foulant detachment (Figure 6.5a). Thus, the foulant could be easily cleaned by water washing. After that, the gates can be recovered to a shrunken state to preserve the permeability (Figure 6.5a). Recently, thermo-responsive surfaces have been shown to reversibly capture and release targeted Michigan Cancer Foundation-7 cells with temperatures changing between 37 and 20°C (Figure 6.5b). This offers opportunities for the gating membranes to be used as smart substrates with a self-cleaning function for cell cultures. Moreover, the negatively K^+-responsive gating membranes can self-clean dead A549 lung carcinoma cells on their surface during the cell culture, due to the swelling of the polymer brush in response to K^+ ions from the dead cells, or in response to a temperature change from 37 to 10°C for dead cell detachment (Figure 6.5c). Such gating membranes with stimuli-induced self-cleaning functions could be a new generation of membranes.

6.3 FUTURE PROSPECTIVE OF pH-RESPONSIVE MEMBRANES

It is convenient to prepare flat sheet pH-sensitive membranes. Thus, most of the studies focused on flat sheet membranes. Because the preparation method and process are different for flat sheet and hollow fiber membranes, the morphology and pH sensitivity are correspondingly slightly different. Due to the large surface/volume ratio, the recent development of pH-sensitive hollow fiber membranes might

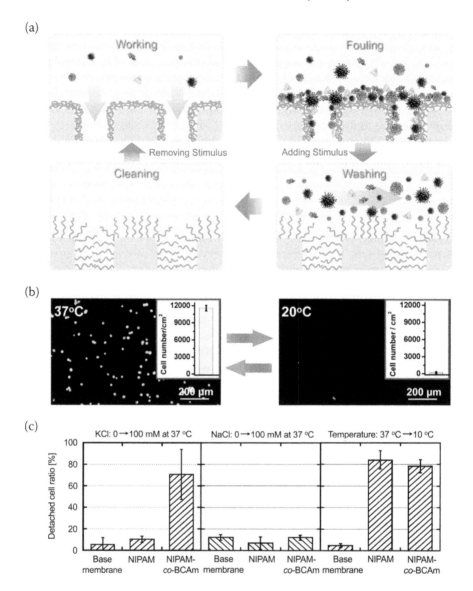

FIGURE 6.5 Smart-gating membranes for self-cleaning. (a) Schematic illustration of the self-cleaning principle with smart-gating membranes through easily adding/removing a simple environmental stimulus, e.g., temperature decrease for PNIPAM gates. (b) Thermo-induced self-cleaning of cells on smart-gating membrane surface. (c) Detached cell ratios of the smart-gating membrane in response to various signals (reproduced with permission from Liu et al. (2016) [28] © Royal Society of Chemistry).

highlight the adaptability of stimuli-responsive membranes to the whole flux control and separation devices [18–20].

To prepare pH-sensitive membranes, polyelectrolytes or weak polyelectrolytes can be used directly, but the strength of the membranes are not always enough for use as separation membranes. These membranes, however, can be used as ion exchange membranes. Today, ion exchange membranes (IEMs) are receiving considerable attention and are successfully used for desalination of sea and brackish water and for treating industrial effluents. They are efficient tools for the concentration or separation of food and pharmaceutical products containing ionic species as well as the manufacturing of basic chemical products [21]. For IEM, a large IEC is needed for binding metal ions. For pH-sensitive membranes, however, small IECs sometimes lead to good pH sensitivity and large water flux change [22,23]. Of course, pH-sensitive membranes with larger IEC could also be used as IEM, and the barrier properties might play their specific function, especially the pH-sensitive hollow fiber membrane devices.

Membrane modification provides a good approach for the preparation of pH-sensitive membranes. The blending method can be used to prepare any kind of membranes, but the matrix polymer should be miscible with the polyelectrolyte or the weak polyelectrolyte. The pore-filled method is extensively investigated to prepare pH-sensitive flat sheet membranes. The flux change can be as large as two orders of magnitude, and MF or UF membranes can be changed to an NF membrane. The grafting method is another good way to prepare pH-sensitive membranes. Both the pore-filled and the grafting methods are allow for easy preparation of flat sheet membranes, but fabrication and modification of hollow fiber membranes can be more difficult. Moreover, the grafted pH-sensitive groups are usually on the outer surface of the hollow fiber membrane wall [24].

pH-responsive membrane systems with changing barrier properties and highly adaptive surfaces have been created using different approaches in recent decades, but most of the work has been focused on the choice of membrane matrix and polyelectrolyte or weak polyelectrolyte, hosting different pH-sensitive groups. In addition, some work has been devoted to understanding how the different responsive interactions within the membranes can be tuned and monitored in controlled environments. The next generation of pH-responsive membranes will move towards advanced functions and beyond barrier functions and will shift from nonspecific triggers to specific, affinity-type triggers [25]. Innovations will lead to the design of more complex membrane systems capable of mimicking functions of living systems, such as pH-sensitive membranes coupled with biocatalytic processes [26].

The modeling of pore-filled pH-sensitive membranes has been performed [27]. Membranes prepared by other methods are seldom studied for separation mechanisms [26]. Thus, studies of the mechanisms for blended membranes and grafting membranes are needed, especially for hollow fiber membranes. Furthermore, the application of pH-sensitive membranes has been widely investigated for controlled drug release, control of water flux, and salt rejection. The real application of pH-sensitive membranes, especially of hollow fiber membrane devices for separation, such as the solute separation, is needed.

REFERENCES

[1] M. Tamada, M. Asano, R. Spohr, J. Vetter, C. Trautmann, M. Yoshida, R. Katakai, and H. Omichi. Preparation of hydrolyzed pH-responsive ion track membrane. *Macromol. Rapid Commun.* 16 (1995) 47–51.

[2] M. S. Oak, T. Kobayashi, H. Y. Wang, T. Fukaya, and N. Fujii. pH effect on molecular size exclusion of polyacrylonitrile ultrafiltration membranes having carboxylic acid groups. *J. Membr. Sci.* 123 (1997) 185–95.

[3] L. F. Gudeman, and N. A. Peppas. Preparation and characterization of pH-sensitive interpenetrating networks of poly(vinyl alcohol) and poly(acrylic acid). *J. Appl. Polym. Sci.* 55 (1995) 919–28.

[4] L. F. Gudeman, and N. A. Peppas. pH-sensitive membranes from poly(vinyl alcohol)/poly(acrylic acid) interpenetrating networks, *J. Membr. Sci.* 107 (1995) 239–48.

[5] S. B. Park, J. O. You, H. Y. Park, S. J. Haam, and W. S. Kim. A novel pH-sensitive membrane from chitosan-TEOS IPN; preparation and its drug permeation characteristics. *Biomaterials* 22 (2001) 323–30.

[6] J. S. Turner, and Y. L. Cheng. Heterogeneous polyelectrolyte gels as stimuli-responsive membranes. *J. Membr. Sci.* 148 (1998) 207–22.

[7] K. Zhang, and X. Y. Wu. Temperature and pH-responsive polymeric composite membranes for controlled delivery of proteins and peptides. *Biomaterials* 25(2004) 5281–91.

[8] Y. Ito, Y. S. Park, and Y. Imanishi. Nanometer-sized channel gating by a self-assembled polypeptide brush. *Langmuir* 16 (2000) 5376–81.

[9] H. Zhang, and Y. Ito. pH control of transport through a porous membrane self-assembled with a poly(acrylic acid) loop brush. *Langmuir* 17 (2001) 8336–40.

[10] A. M. Hollman, and D. Bhattacharyya. Controlled permeability and ion exclusion in microporous membranes functionalized with poly(l-glutamic acid). *Langmuir* 18 (2002) 5946–52.

[11] Y. Ito, S. Kotera, M. Inaba, K. Kono, and Y. Imanishi. Control of pore size of polycarbonate membrane with straight pores by poly(acrylic acid) grafts. *Polymer* 31 (1990) 2157–61.

[12] Y. Ito, M. Inaba, D. Chung, and Y. Imanishi. Control of water permeation by pHand ionic strength through a porous membrane having poly(carboxylic acid) surface-grafted. *Macromolecules* 25 (1992) 7313–16.

[13] A. M. Mika, R. F. Childs, J. M. Dickson, B. E. McCarry, and D. R. Gagnon. A new class of polyelectrolyte-filled microfiltration membranes with environmentally controlled porosity. *J. Membr. Sci.* 108 (1995) 37–56.

[14] T. Peng, and Y. L. Cheng. pH-responsive permeability of PE-g-PMAA membranes, *J.Appl. Polym. Sci.* 76 (2000) 778–86.

[15] Y. Wang, Z. Liu, B. Han, Z. Dong, J. Wang, D. Sun, Y. Huang, and G. Chen. pH sensitive polypropylene porous membrane prepared by grafting acrylic acid in-supercritical carbon dioxide. *Polymer* 45 (2004) 855–60.

[16] L. Y. Chu, T. Niitsuma, T. Yamaguchi, and S. Nakao. Thermo-responsive transport through porous membranes with grafted PNIPAM gates. *AIChE J.* 49 (2003) 896–909.

[17] M. Yang, L. Y. Chu, H. D. Wang, R. Xie, H. Song, and C. H. Niu. A thermo-responsive membrane for chiral resolution. *Adv. Funct. Mater.* 18 (2008) 652–63.

[18] B. S. Qian, J. Li, Q. Wei, P. L. Bai, B. H. Fang, and C. S. Zhao. Preparation and characterization of pH-sensitive polyethersulfone hollow fiber membrane for flux control. *J. Membr. Sci.* 344 (2009) 297–303.

[19] T. Xiang, Q. H. Zhou, K. Li, L. L. Li, F. F. Su, B. S. Qian, and C. S. Zhao. Poly (acrylic acid-co-acrylonitrile) copolymer modified polyethersulfone hollow fiber membrane with pH-sensitivity. *Sep. Sci. Technol.* 45 (2010) 2017–27.

[20] S. Suryanarayan, A. M. Mika, and R. F. Childs. Gel-filled hollow fiber membranes for water softening. *J. Membr. Sci.* 281 (2006) 397–409.

[21] T. W. Xu. Ion exchange membranes: state of their development and perspective. *J. Membr. Sci.* 263 (2005) 1–29.

[22] Q. Wei, J. Li, B. S. Qian, B. H. Fang, and C. S. Zhao. Preparation, characterization and application of functional polyethersulfone membranes blended with poly(acrylic acid) gels. *J. Membr. Sci.* 337 (2009) 266–73.

[23] W. Zou, Y. Huang, J. Luo, J. Liu, and C. S. Zhao. Poly(methylmethacrylate–acrylic acid–vinylpyrrolidone) terpolymer modified polyethersulfone hollow fiber membrane with pH-sensitivity and protein antifouling property. *J. Membr. Sci.* 358 (2010) 76–84.

[24] S. Bequet, J. C. Remigy, J. C. Rouch, J. M. Espenan, M. Clifton, and P. Aptel. From ultrafiltration to nanofiltration hollow fiber membranes: a continuous UV-photografting process. *Desalination* 144 (2002) 9–14.

[25] D. Wandera, S. R. Wickramasinghe, and S. M. Husson. Stimuli-responsive membranes. *J. Membr. Sci.* 357 (2010) 6–35.

[26] X. R. Chen, Z. G. Su, G. H. Ma, and Y. H. Wan. Studies of intelligent membranes for separation. *Prog. Chem.* 18 (2006) 1218–26.

[27] K. Hu, and J. M. Dickson. Modelling of the pore structure variation with pHfor pore-filled pH-sensitive poly (vinylidene fluoride)–poly(acrylic acid) membranes. *J. Membr. Sci.* 321 (2008) 162–71.

[28] Z. Liu, W. Wang, R. Xie, X. -J. Ju and L. -Y. Chu. Stimuli-responsive smart gating membranes. *Chem. Soc. Rev.* 45 (2016) 460–75.

Index

For Product Safety Concerns and Information please contact our EU
representative GPSR@taylorandfrancis.com
Taylor & Francis Verlag GmbH, Kaufingerstraße 24, 80331 München, Germany

www.ingramcontent.com/pod-product-compliance
Ingram Content Group UK Ltd.
Pitfield, Milton Keynes, MK11 3LW, UK
UKHW021123180425
457613UK00005B/201